Microelectronics

Maurizio Di Paolo Emilio

Microelectronics

From Fundamentals to Applied Design

Maurizio Di Paolo Emilio
EDM Engineering
Pescara, Italy

ISBN 978-3-319-36423-0 ISBN 978-3-319-22545-6 (eBook)
DOI 10.1007/978-3-319-22545-6

Springer Cham Heidelberg New York Dordrecht London

Softcover re-print of the Hardcover 1st edition 2016

Printed on acid-free paper

Springer International Publishing AG Switzerland is part of Springer Science+Business Media (www.springer.com)

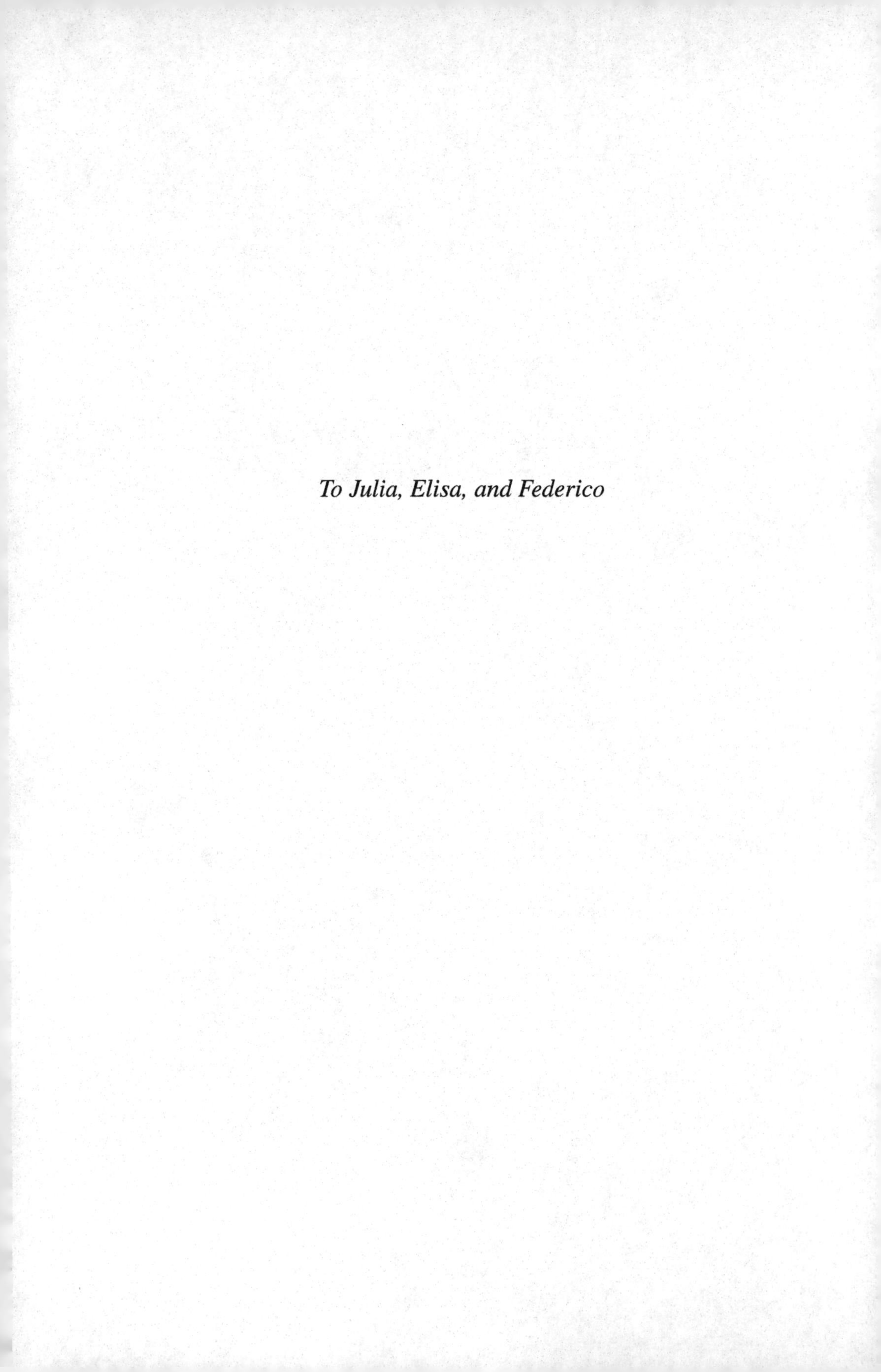

To Julia, Elisa, and Federico

Imagination is more important than knowledge [A. Einstein].

When wireless is perfectly applied, the whole earth will be converted into a huge brain, capable of response in every one of its parts [N. Tesla].

Preface

Microelectronics is related to the study and manufacture of tiny electronic designs and components. These devices are typically made from semiconductor materials and named integrated circuits. An integrated circuit or monolithic integrated circuit is a set of electronic circuits on one small plate of semiconductor material, normally silicon. This can be made much smaller than a discrete circuit made from independent components. ICs can be made very compact, and have up to several billion transistors and other electronic components in an area the size of a fingernail.

Microelectronics such as microprocessors and microcomputers are capable of performing complex data processing tasks. To be able to take advantage in the microelectronic technology, there will always be a need for suitably qualified engineers with experience and know-how of the latest technological developments and the tools to support them.

Moreover, a typical PCB contains a large number of electronic components, most of which are tiny in size. Without it, it would be nearly impossible to connect all components together with wires. A PCB provides a convenient platform to arrange the electronic components in a compact and efficient way. This compactness allows the deep development and complicated electronic circuits in small form factors, taking less space in devices.

The electronics was founded in 1907 with the invention of the triode, the first vacuum tube, which can amplify an electrical signal. For the first few decades of life, the essential function of the valves was the amplification of signals, allowing the development of wireless telegraphy, radio, amplifier, and television. During World War II, the range of applications of the valves is extended to the radar, control systems, and electronic calculation, culminating in Colossus (1944), the electronic calculator, which was developed by British scientists and specialized to decode secret messages between Hitler and his high command. In these systems, the valve was used as a switch, instead of amplifier, working on binary signals, by providing a much faster to electromechanical relays. The search to replace the valve with a smaller device, less expensive and more energy efficient, which began in 1930, finally led to the invention of the transistor in 1947 by three scientists at Bell Labs. Thus, microelectronics was born. For the first 10 years, the transistors were built one at a time using germanium as a semiconductor element. In 1959, the invention of the planar process to Fairchild Semiconductor, at the hands of the Swiss Jean Hoerni, changed everything. It thus began to build transistors, a hundred at a time, on a silicon wafer—element semiconductor with more favorable characteristics of germanium—by removing the germanium in a few years. The advent of the integrated circuit began to replace not only the active components but also the passive ones: resistors, capacitors, diodes, etc., as well as the portion of the printed circuit, which is necessary to link them together. Further fundamental stage of microelectronics was the invention of a new type of transistor, the MOS transistor (Metal Oxide Semiconductor), which once perfectly began to replace the bipolar transistors used in all of the first integrated circuits. With the invention of MOS process, it was possible to build the first semiconductor memories and the first microprocessor. The application field of microelectronics widened again, replacing the memories to magnetic ferrite used in computers and creating for the first time an entire computer in the solid state. Within another decade, MOS circuits are eventually supplant for almost all bipolar chips, simultaneously extending their application capabilities to analog ICs and the nonvolatile memory and image sensors; the last two applications impossible to do both with bipolar transistors and with MOS transistors with gate aluminum. In 1965, Gordon Moore, one of the two founders of the Intel (1968), then head of the research laboratories of Fairchild Semiconductor, said that each year the number of transistors on a chip doubled, thinking that this behavior would be continued in the future. This observation was later called Moore's Law, although obviously not a physical law as the law of Newton and its validity is limited in time. The longevity of Moore's law is due to the principle of scaling, discovered a few years later, according to which, by reducing the critical dimensions of the MOS transistor in the same proportion, not only decreases the area of the transistor and its power loss, but also increases its speed. Since in the mid-seventies, scaling has been the key strategy to improve the performance and reduce the cost of integrated circuits. In 1970, the most advanced lithography in production was able to print lines 6 μm. Today, we got less to 45 nm, by reducing the area of a transistor about 20,000 times in 38 years. In 1970, the RAM was static and had more advanced 256-bit, with an access time of 1.5 μs. Today, the most advanced static RAM has more than 64 Mbit with an access time of 2.6 ns. Still more rapid has performed in the progress in the field of dynamic RAM memories and flash memories, which did not yet exist in the flash memory in 1970. The most advanced in the production today has 512 million cells, with an access time of about 50 ns, integrating about 1.6 billion transistors. With the flash memory, it is even possible to store more than 1 bit per cell, measuring the number of electrons that are injected into the floating gate of the memory cell and reaching up to 8 Gbit memory with a chip 512 million cells. In the field of microprocessors progress, it was equally amazing. The first microprocessor, the Intel 4004, introduced in the market in 1971, incorporated about 2300 transistors in an area of 12 mm square. The 4004 was able to run about 100,000 instructions per second, operating on 4 bits. Each instruction cycle required eight clock periods, at a frequency of about 750 kHz. The 4004 had 16 pins, the working voltage was 15 V, and the power loss was 750 mW. If we compare the 4004 with one of the latest Intel microprocessors, the Xeon L5420, introduced earlier this year, we can see the extraordinary progress made in 37 years. The Xeon integrates about 820 million transistors with 45 nm, occupies an area of 210 mm square, and is manufactured using 30 masks. The Xeon has four CPUs working in parallel, which runs up to 20 billion instructions per second, each operating on 64-bit. The clock frequency is 2.5 GHz. The Xeon has 771 pins, dissipates 50 W, and uses a working voltage of 1.2 V. In the year of 2000, Intel announced a new series of Xeon microprocessors, the 7400 series, with 6 cores and 16 MB of cache memory, integrating well 1.9 billion transistors on the same chip. In the history of technological progress, microelectronics has far exceeded any other discipline in increasing performance over time, maintaining the same cost. Intel has succeeded in developing a stack proper, after decades of effort, bringing into production with first chips to 45 nm, 22 nm, and so on. It will be possible to have 10 nm chips in production in the next years. The construction process to 10 nm (10 nm), initially 11 nm according to the International Technology Roadmap for Semiconductors, is the evolution of the future process to 14 nm used for microprocessors from Intel and AMD (over that for other types of circuits made from other companies in the sector) and its introduction is planned between 2015 and 2016. The correct name for this stage of the technology comes from the International Technology Roadmap for Semiconductors (ITRS), which, however, in its roadmap had marked as future technologies the threshold of 16 nm in 2015 (later updated to 14 nm in 2013 and 11 nm in 2022 (later updated to 10 nm in 2015).

The term "10 nm" indicates the average size of the gate of each transistor. To get an idea of what "10 nm" is, it is enough to consider that the HIV is about 120 nm, a human red blood cell is approximately 6000–8000 nm, and a human hair is about 80,000 nm. The advantages in the switch to this constructive process and, more generally, to try to improve more and more miniaturization are varied, ranging from the improvement of the production yield, with consequent reduction of costs (more than one processor is "small" and multiple processors may be manufactured with only one wafer), the decrease of the electricity consumption, through the possibility of integrating an ever greater number of transistors with a consequent increase in processing power. An evolutionary strategy that is already maturing before our eyes is to make chips with more than one active layer. The more elegant method, but also more complex, is to build the chip with more than one active layer on the same slice of silicon. For example, the Foveon has built image sensors where the three photodiodes for the detection of the primary colors are one above the others rather than next to one another. The most practical method, and perhaps at the end also less expensive, however, to mount more than one chip, one above the other, on the same package. Already most of the flash memories using these techniques to increase the number of bits of memory contained in the same device. But the road will be based on the most advanced new techniques which now goes under the generic name of wafer-scale-packaging. WLSC Methods, which are able to extend the same principle used in the planar process, are also encapsulation of the chips, by assembling and then all at once the chips that are in a silicon wafer, and separating at the end of the process, ready to be mounted on printed circuit boards. All this activity is creating the necessary knowledge to start a new road that goes by the generic name of 3D ICs, chips in three dimensions. A "three-dimensional integrated circuit" (3D IC) is an integrated circuit manufactured by stacking silicon wafers and/or dies and interconnecting them vertically using through-silicon vias (TSVs) so that they behave as a single device to achieve performance improvements to reduce power and smaller footprint than conventional two-dimensional processes. They can be classified by their level of interconnect hierarchy at the global (package), intermediate (bond pad), and local (transistor) level. In general, 3D integration is a broad term that includes such technologies as 3D wafer-level packaging (3DWLP). There is no doubt that in the future 3D ICs, which today are only the beginning of their development, they will receive more and more attention and become the common use gradually scaling that will become increasingly difficult and expensive, thus prolonging the period of the continuously improvements to the microelectronics even after scaling traditional has finished its course. Use the term "microelectronics" even when the size of transistors is a few tens of nm, when we could very well use the term "nanoelectronics." This is because I would like to reserve the term "nano" to a new class of electronic devices, smaller and faster, based on new principles of operation, which promise to replace the MOS transistors. Nanoelectronics therefore offers the new road to continually increase the performance and reduce the size and cost of the integrated circuits once the MOS transistors have reached the physical limit of scaling. Nanoelectronics is now more than a science technology. In fact it has not yet reached the level of commercialization, although progress over the past decade has been phenomenal. At the nanometer scale, electrical and magnetic properties are surprising in both old and new materials, giving rise to many new lines of research.

The breathtaking advances in the field of microelectronics over the past 20 years have made the implementations and realizations of real-time fast and power-efficient computer systems, digital signal processing (DSP) systems, communication systems, biomedical systems, and systems for consumer goods a reality. Through the study of this text, the reader will develop a comprehensive understanding of the basic techniques of modern electronic circuit design and techniques of PCB design.

Printed circuit boards or PCBs have become an integral part of electronic equipment. A typical PCB contains a large number of active and passive components connected together through traces on the board and are helpful in performing diagnostics for a number of reasons.

Even though, most readers may not ultimately be engaged in the design of integrated circuits (ICs) themselves, a thorough understanding of the internal circuit structure of ICs is prerequisite to avoiding many pitfalls that prevent the effective and reliable application of integrated circuits in system design.

The main idea of this book is to describe the fundamental theory of Microelectronics starting with PN junction until operational amplifier. Some aspects of PCB design and recent technology for application as sensors and control, and data acquisition systems will be analyzed.

Starting from the review of analogue electronics, the book aims to provide the reader into a competent and independent practitioner in the field of Microelectronics by providing several skills.

In particular, will be analyzed BJT and MOS transistor with its application. Then, application with operational amplifier and some consideration about the recent technologies: 4D electronics, CMOS Sensor, GaN material, and so on.

Pescara, Italy
June 2015

Maurizio Di Paolo Emilio

Acknowledgments

I would like to express my gratitude to all those who gave me the possibility to complete this book. In particular, I want to thank Charles B. Glaser, Senior Editor—Electrical Engineering, for the publication of the present book. To my family, thank you for patience, encouraging me, and inspiring me to follow my dreams. I am especially grateful to my wife, Julia, and my children, Elisa and Federico.

Contents

Chapter 1
Review of Microelectronics

Abstract Microelectronics is become more useful in the embedded system in particular in the Data Acquisition, thanks to the development of the semiconductor technology. The goal of this chapter is to review briefly the main features of electronics and microelectronics; one salient feature of this review is its synthesis and design-oriented approach. In the next chapters we will analyze in detail the main specifications.

1.1 Introduction

The earliest electronic circuits were fairly simple. They were composed of a few tubes, transformers, resistors, capacitors, and wiring. As more was learned by designers, they began to increase both the size and complexity of circuits. Component limitations were soon identified as this technology developed. The transition from vacuum tubes to solid-state devices took place rapidly. As new types of transistors and diodes were created, they were adapted to circuits. The reductions in size, weight, and power use were impressive.

Microelectronic technology today includes thin film, thick film, hybrid, and integrated circuits and combinations of these. Such circuits are applied in DIGITAL, SWITCHING, and LINEAR (analog) circuits. Because of the current trend of producing a number of circuits on a single chip, you may look for further increases in the packaging density of electronic circuits. At the same time you may expect a reduction in the size, weight, and number of connections in individual systems. Improvements in reliability and system capability are also to be expected [1, 2].

1.2 Basics of Semiconductor's Physics

Microelectronics is based on physics of structures. The main semiconductor elements used in the applications are silicon, boron, aluminum, etc. Silicon atom has 4 valence electrons and it is important for electrical system. At temperature close to 0 K, silicon crystal behaves as an insulator: its electrons are confined to their respective covalent bonds. At higher temperature, electrons gain thermal energy and can be used as "free charge." Bandgap energy, minimum energy to break the bond,

M. Di Paolo Emilio, *Microelectronics*, DOI 10.1007/978-3-319-22545-6_1

of the silicon is 1,12 eV where 1 eV $= 1.6 * 10^{-19}$ J. The number of "free electrons" that is possible to have in a semiconductor crystal depends on band gap energy and temperature, it's possible to define density of electrons (number of electrons per unit volume) as the following [3]:

$$n = 5.2 * 10^{15} T^{\frac{3}{2}} * e^{-\frac{E_g}{2KT}} \tag{1.1}$$

with k = Boltzmann constant = 1.3810^{-23} J/K. For insulator materials $E_g \simeq$ 2.5 eV, instead, for conductor is about less than 1 eV. As example we can calculate the density of electrons for silicon at $T = 300$ K:

$$n = 5.2 * 10^{15} T^{\frac{3}{2}} * e^{-\frac{1.72*10^{-19}}{2KT}} \sim 10^{10}\,\text{e/cm}^3 \tag{1.2}$$

The intrinsic semiconductors can't be used for electrical application due to not only very high resistance but also a small number of "free electrons" with respect to the conductor. In an extrinsic semiconductor the situation is different: there is a p-Si and an n-Si. In the first case majority carriers (holes) are p $= N_A$, semiconductor is doped with a density of N_A (i.e., boron atom) and minority carriers $n = p_i^2/N_A$; in the second case, instead, majority carriers (electrons) n $= N_A$ with semiconductor is doped with a density of N_D (i.e., phosphorus), minority carriers $p = n_i^2/N_D$. The concept of doping extends the flow of current in a semiconductor.
The drift is the movement of electrons charge due to an electric field: $v = \mu E$, μ is "mobility" expressed in cm^2/V s. Considering a piece of n-silicon ($\mu_n =$ 1350 cm^2/V s) of $L = 1\,\mu$m applied a voltage of $V = 1$ V, the velocity of electrons (drift) is about $v = \mu_n E = \mu_n \frac{V}{L} \simeq 10^7$ cm/s. For a p-silicon $\mu_p = 480$ cm^2/s [1, 2].

Now, we should calculate the current due to this drift v (Fig. 1.1):

$$I = -v \cdot W \cdot h \cdot n \cdot q \tag{1.3}$$

It can be written in another way:

$$J_n = \mu_n \cdot E \cdot n \cdot q \tag{1.4}$$

where J is the current density, $W \cdot h$ is the cross section, and $n \cdot q$ is the charge density in Column of the semiconductor. In presence of both electrons and holes:

$$J = J_n + J_p = q(\mu_n n + \mu_p p)E \tag{1.5}$$

The mechanism of the diffusion is the movement of the charge from a zone of high concentration to zone of low concentration (current). The mathematical relation that explains this phenomena is the following:

$$I = A \cdot q \cdot D \cdot n \frac{dn}{dx} \tag{1.6}$$

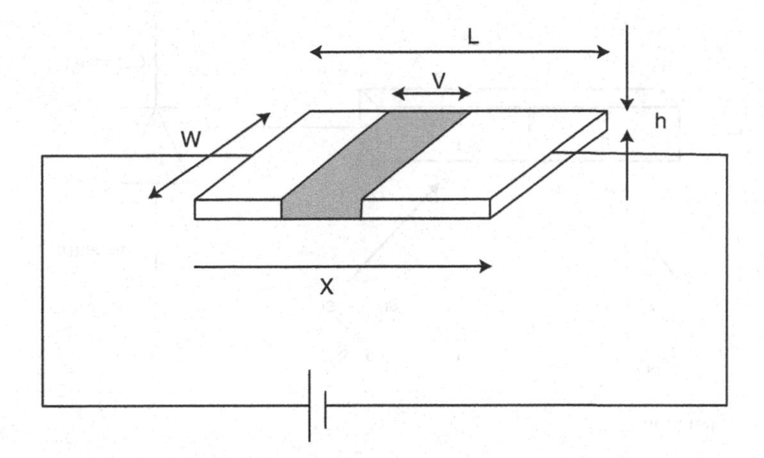

Fig. 1.1 Current flow in terms of charge density

where D_n is the "diffusion constant" in cm^2/s; in intrinsic silicon, $D_n = 34\,cm^2/s$ (for electrons) and $D_p = 12\,cm^2/s$ (for holes). A is the cross section, q is the charge, and $\frac{dn}{dx}$ is the concentration (electrons) with respect to the direction "x" (Fig. 1.1). From the last equation we can calculate the current density for electrons and holes:

$$J_n = qD_n \frac{dn}{dx} \tag{1.7}$$

$$J_p = -qD_p \frac{dp}{dx} \tag{1.8}$$

1.2.1 PN Junction

From the doping we have obtained p-type and n-type semiconductors. An electric field or a concentration gradient leads to the movement of the charges (electrons and holes). We suppose to dope two adjacent pieces of semiconductors (Fig. 1.2); in this configuration, there will be a flow of electrons from n to p side and a flow of holes in opposite direction (Fig. 1.3). At the end of this process of "equilibrium," an electric field (depletion region) will emerge as indicated in Fig. 1.3. The Junction reaches equilibrium once the electric field is strong enough to completely stop the diffusion current. The existence of electric field due to depletion region suggests that PN junction has a built-in potential defined in the following equation:

$$V(x_2) - V(x_1) = -\frac{D_p}{\mu_p} \ln\left(\frac{p_p}{p_n}\right) \tag{1.9}$$

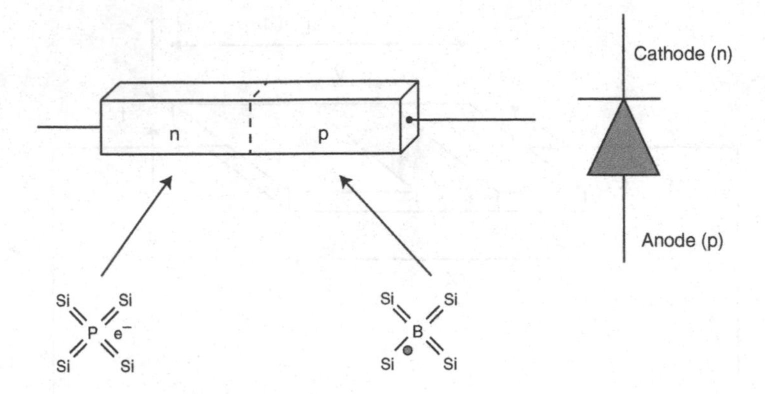

Fig. 1.2 PN junction

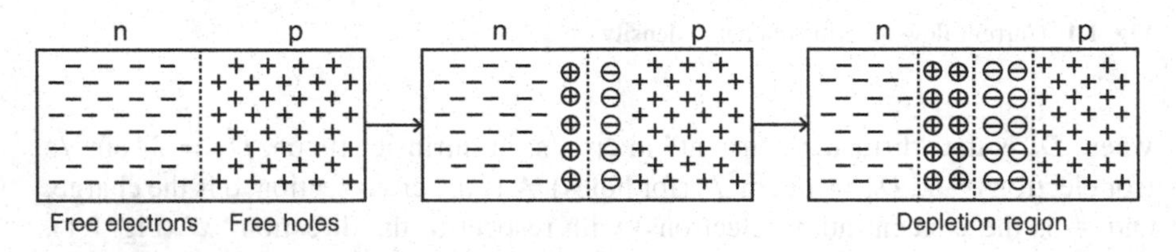

Fig. 1.3 Evolution of charge concentration in a PN junction

Considering the Einstein's equation, $\frac{D}{\mu} = \frac{KT}{q}$:

$$|V_0| = \frac{KT}{q} \ln\left(\frac{p_p}{p_n}\right) \tag{1.10}$$

where p_n and p_p are the concentrations at x_1 and x_2, respectively. The PN junction so realized is called Diode. Having analyzed the PN junction in equilibrium, let us observe how it works applying an external voltage (Fig. 1.4):

The external voltage rises the electric field of depletion region prohibiting the flow of current. The device works as capacitor:

$$C_j = \frac{C_{j0}}{\sqrt{1 - \frac{V_R}{V_0}}}; C_{j0} = \sqrt{\frac{q\epsilon_{s_i}q}{2} \frac{N_A N_D}{N_A + N_D} \frac{1}{V_0}} \tag{1.11}$$

ϵ_{s_i} is the dielectric constant of silicon.

The external voltage decreases the electric field of the depletion area allowing greater diffusion current (Fig. 1.4):

$$I_D = I_s \left(\exp\left(\frac{V_R}{V_T}\right) - 1\right) \tag{1.12}$$

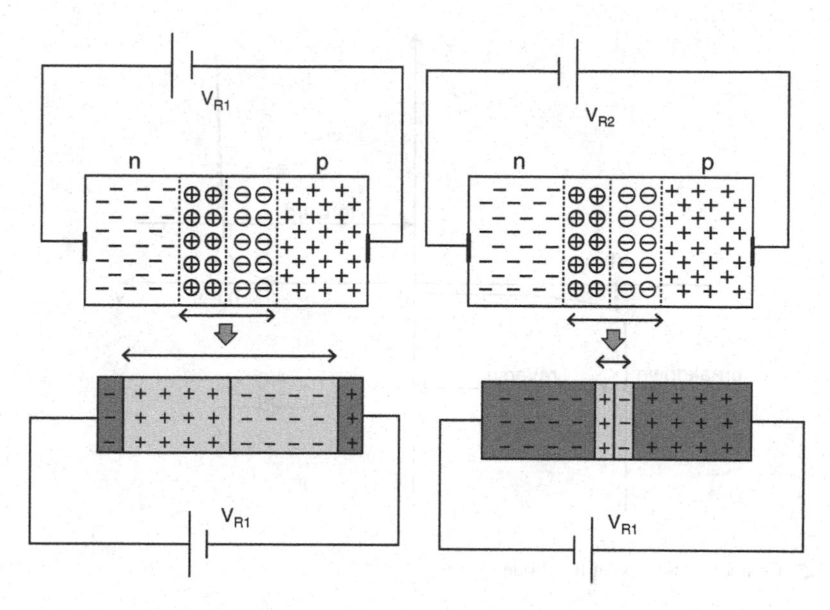

Fig. 1.4 PN junction with external voltage

with V_T, thermal voltage and I_s:

$$I_s = Aqn_i^2 \left(\frac{D_n}{N_A L_n} + \frac{D_p}{N_D L_p} \right) \tag{1.13}$$

is the "reverse saturation current" and L_n and L_p are the electrons and holes "diffusion lengths" (i.e., $L_n = 20\,\mu\text{m}$, $L_p = 30\,\mu\text{m}$)

1.3 Diode

The diode is a two-terminal device with I–V characteristic indicated in Fig. 1.5. Some application can be described in the following text [1]:

- Wave rectifier: the circuit is visualized in Fig. 1.6: the ripple amplitude can be calculated by:

$$V_r \sim \frac{V_p - V_{D,\text{on}}}{R_L C_1 f_{\text{in}}} \tag{1.14}$$

- Voltage regulation: Sometimes, it is necessary to have, in output of the diode, a stabilized voltage to reduce the significant variation of the line voltage. A possible outline is visualized in Fig. 1.7.

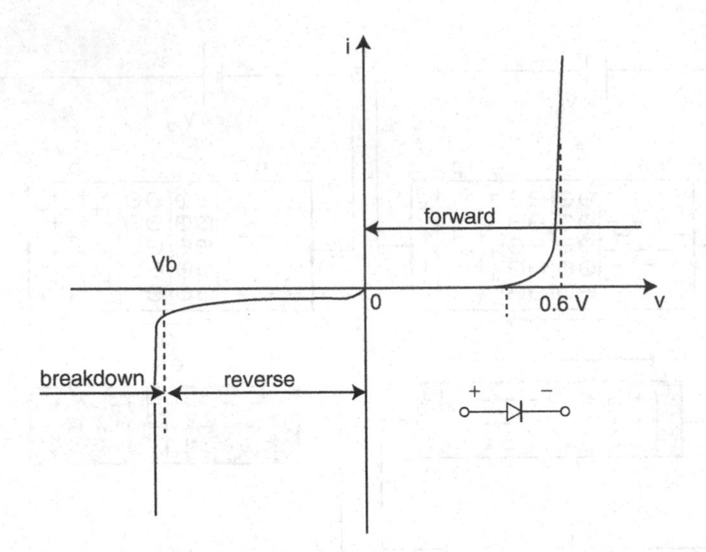

Fig. 1.5 Characteristic I–V of the diode

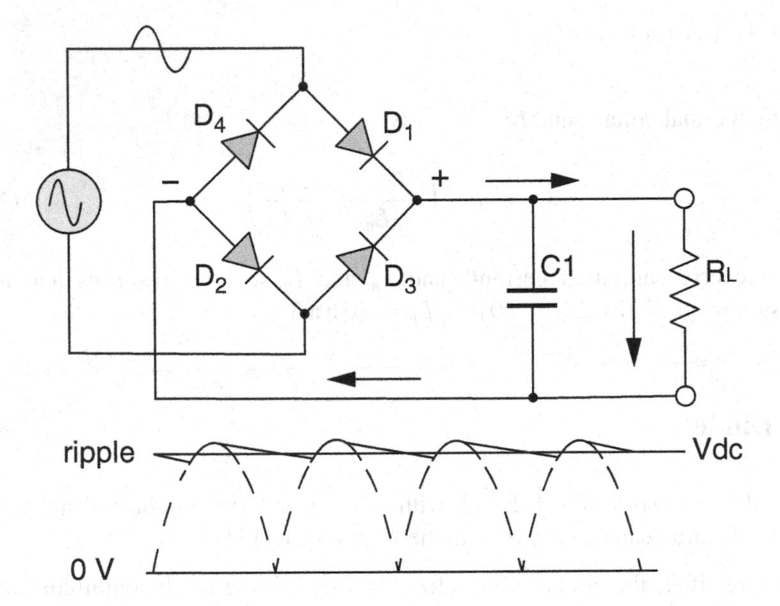

Fig. 1.6 Wave rectifier

- Limiting circuit: The circuit passes the input to the output, $V_{out} = V_{in}$ and when the input exceeds a "threshold" the output remains constant (Fig. 1.8).

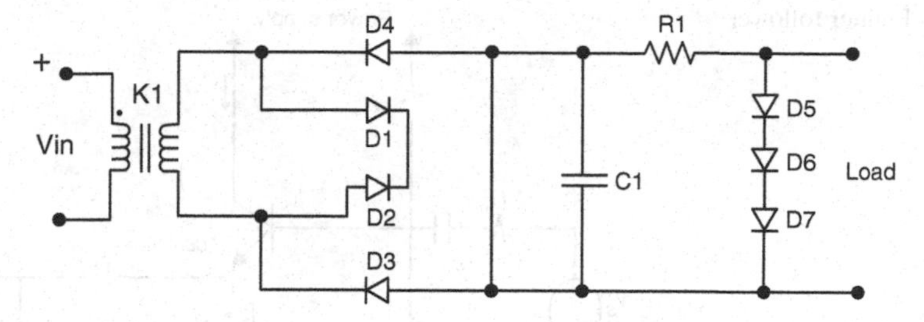

Fig. 1.7 Block diagram of voltage regulator with diodes

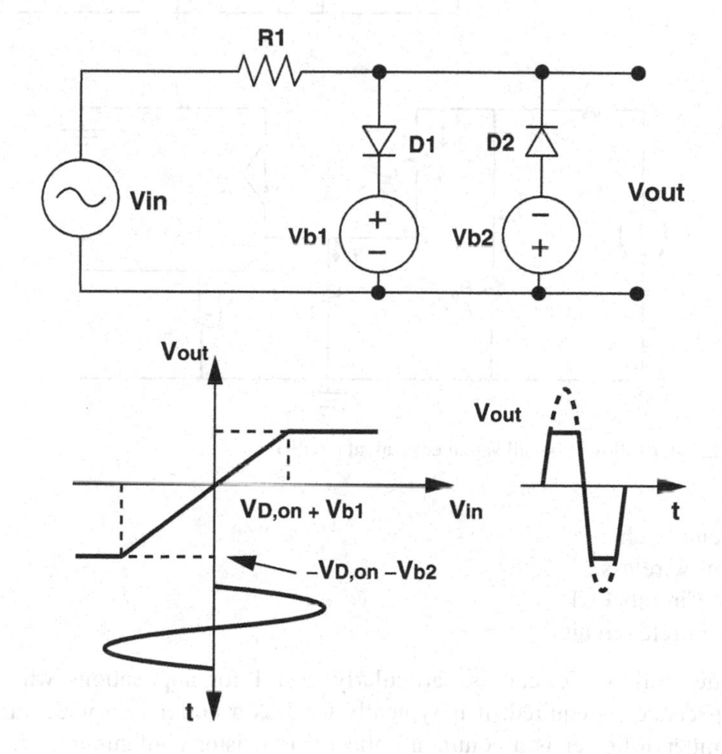

Fig. 1.8 Limiting circuit

1.4 Bipolar Transistor: Emitter Follower

The union of two junctions p-n (i.e., two diodes together) forms the bipolar transistor junction (BJT). Bipolar because the current is sustained by electrons and holes (such as the diode). Compared to the diode, the BJT (three terminals) can be used as a signal amplifier. Although the MOS technology (see next paragraph) is more widespread, the technology bipolar remains significant (or predominant, in certain cases) in several applications:

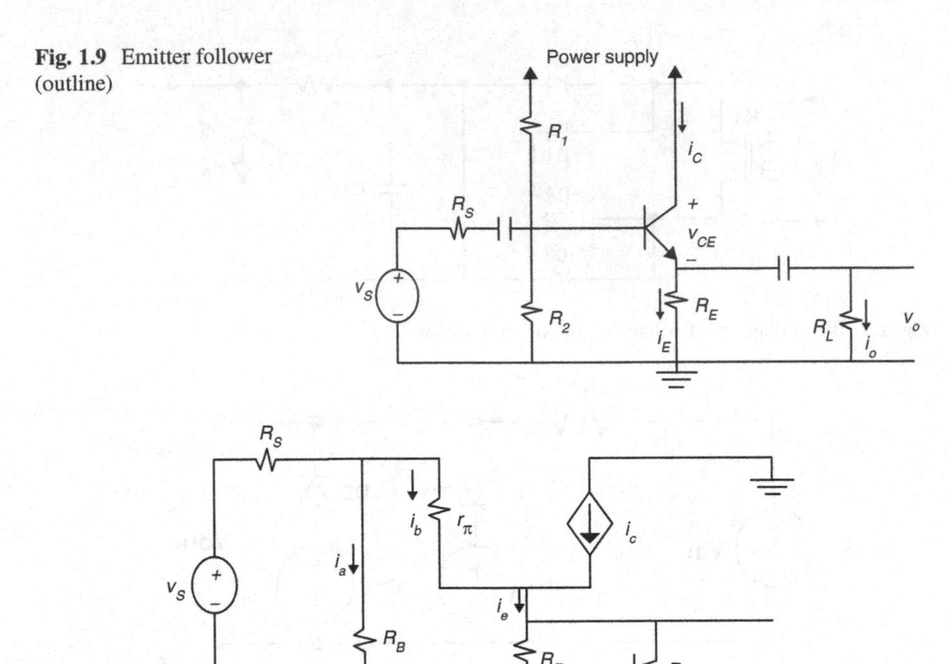

Fig. 1.9 Emitter follower (outline)

Fig. 1.10 Emitter follower (small signal equivalent circuit)

- Electronics vehicle
- Systems wireless
- Digital Circuits ECL
- Draft discrete circuits

The emitter follower circuit is particularly useful for applications where high input impedance is required. It is typically used as a buffer in a wide variety of areas. Emitter follower is a common collector transistor configuration. It can be easily designed by circuit RC. The emitter follower is also known as a voltage follower, or a negative current feedback circuit, with high input impedance and low output impedance. The outline of the emitter follower is shown in Fig. 1.9 and the corresponding equivalent small signal circuit in Fig. 1.10. We can calculate the effective resistance seen from terminal B by:

$$R_{ib} = r_\pi + (\beta + 1)R_E//R_L \simeq (\beta + 1)R_E//R_L \tag{1.15}$$

This value is relatively high. In general, $R_E//R_L$ is around kΩ and $\beta \sim 100$, so the resistance seen from terminal B is in the hundreds of kΩ. It is evident from Fig. 1.10 that input resistance depends on load resistance. Let us name the input resistance as R_{in}:

$$R_{\text{in}} = R_B//R_{ib} \tag{1.16}$$

The effective resistance is the parallel between R_{ib} and R_b. A common-collector configuration can be used as an amplifier in such a circuit, where a large input resistance is needed; a good application can be a pre-amplifier circuit. From the voltage gain equation ($A = V_0/V_i$) follows:

$$A = \frac{V_0}{V_s} = \frac{R_E//R_L}{r_e + (R_E//R_L)} * \frac{R_{\text{in}}}{R_s + R_{\text{in}}} \quad (1.17)$$

It becomes evident that as long as $r_e \ll (R_E//R_L)$ and $R_s \ll R_{\text{in}}$ the gain approaches unity.

The emitter follower can be designed using the main steps described below:

- Choose a transistor: the transistor should be selected according to the system requirements.
- Select emitter resistor: selecting a working point (for example, select an emitter voltage of about half the supply voltage).
- Determine the base current: base current is the collector current divided by β (or h_{fe}).
- Determine the base resistor values: select the value of the resistor(s) to provide the voltage required at the base.
- Determine the value of the input and output capacitor: The value of the input/output capacitor should equal to the resistance of the input/output circuit at the lowest frequency of operation.

When using the emitter follower circuit, there are two main practical points to note:

- The collector may need decoupling: in some cases, the emitter follower may oscillate, in particular if long leads are present. One of the easier ways to prevent this is to decouple the collector to ground with very short connections, or by placing a small resistance between the collector and the power supply line.
- The input capacitance affects the RF: the base emitter capacitance may reduce the high impedance of the input circuit if the signal is above 100 kHz.

One of the designs of an emitter follower is shown in Fig. 1.11, with a P-spice simulation in Figs. 1.12 and 1.13. This configuration can be typically used in AC coupling applications. It's important that the frequency response and bias voltage on the base are planned for. In the circuit mentioned here, there's an added voltage divider, consisting of R_2 and R_3, and an AC coupling capacitor, C_1. Their component values will need to be calculated. An important aspect of the emitter follower is its relative immunity to temperature instability. When the current in the collector increases (changing of β), the voltage across the emitter resistor also increases. This acts as a negative feedback, since it reduces the voltage difference between base and emitter (V_{BE}), which drives the transistor to conduct more current, as such thermal stability is maintained.

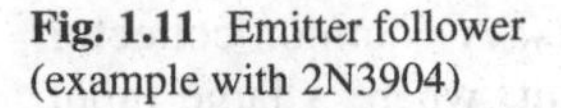
Fig. 1.11 Emitter follower (example with 2N3904)

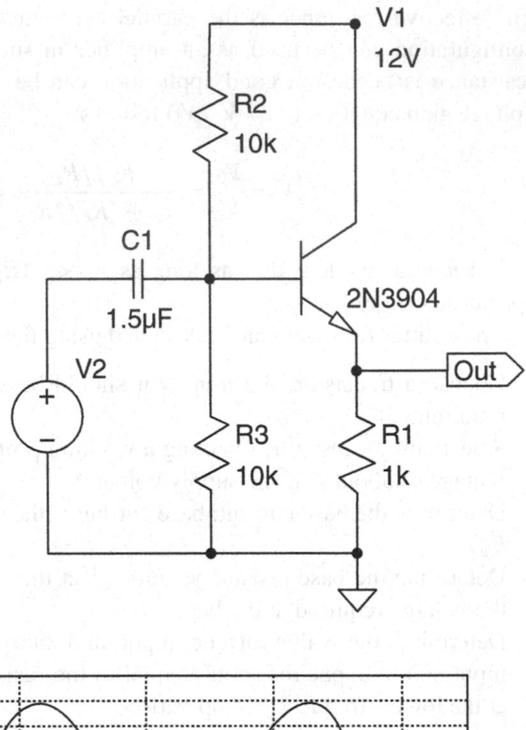

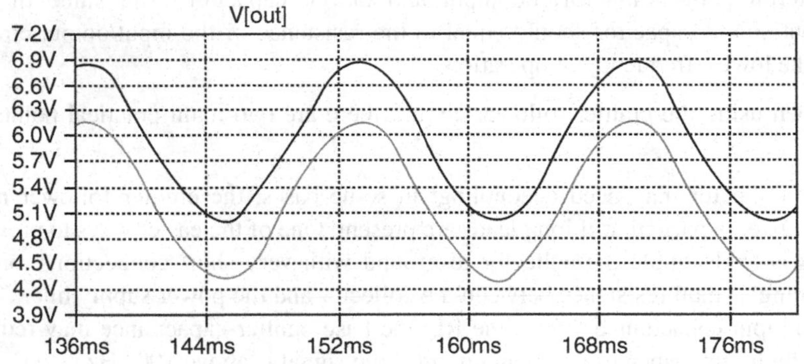

Fig. 1.12 Simulation of emitter follower, output voltage

1.5 MOS Transistor

Today the microelectronics is dominated by MOSFET devices. The physical structure of the MOSFET is described in Fig. 1.14: on a substrate of monocrystalline p-type, two junctions are made of type n+ which are connected to two terminals called drain and source (D and S in Fig. 1.14). In the area between the drain and source is made to grow a layer of silicon dioxide (thickness less than 0.01 µm), and that is an excellent insulator [3–5].

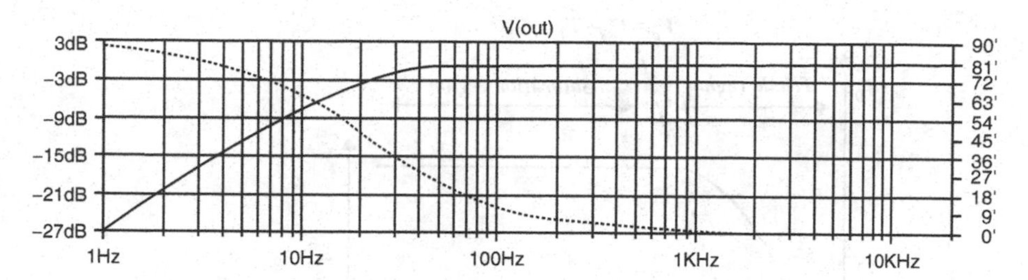

Fig. 1.13 Simulation of emitter follower, frequency response

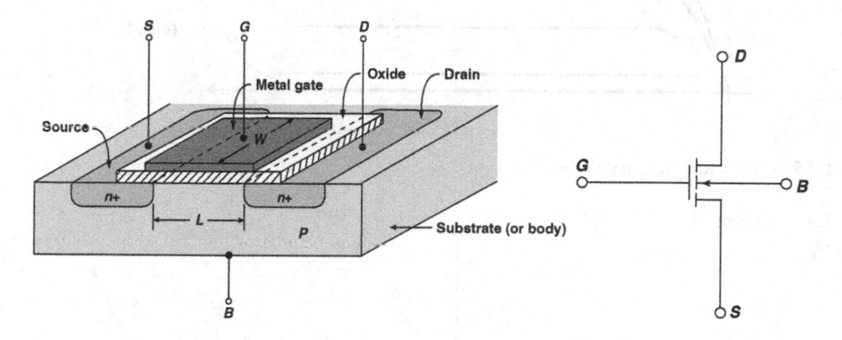

Fig. 1.14 Inside outline (general) and symbol of Mosfet

The gate terminal is isolated from the Si substrate by a thin layer of SiO_2, and therefore the gate current DC is null. The body terminal is generally connected to that source (only three terminals). Drain and Source are symmetrical.

Applying a sufficiently positive voltage between the gate and source, the electrons are attracted towards the Si–SiO_2 interface under the gate forming a conductive channel between source and drain. In these conditions, if it is applied a $V_{ds} > 0$, a current may flow between the drain and source. The current between drain and source is controlled by the voltage between gate and source, which controls the formation of the channel. The electrical characteristics of the MOSFET depend on L (gate length) and W (gate width), as well as technological parameters such as oxide thickness and doping of body. Typical values of L and W are: $L = 0.1$–$2\ \mu\text{m}$, $W = 0.5$–$500\ \mu\text{m}$. The range of gate oxide thickness is 3–50 nm.

There are four types of MOS transistors: two to channel n and two to channel p. The MOSFET n-channel (nMOS) is formed on a p-type substrate:

- nMOS enrichment (enhancement) or normally off
- nMOS depletion (depletion) or normally on

The MOSFET p-channel (pMOS) is formed on a substrate of n-type:

- pMOS enrichment (enhancement) or normally off
- pMOS depletion (depletion) or normally on

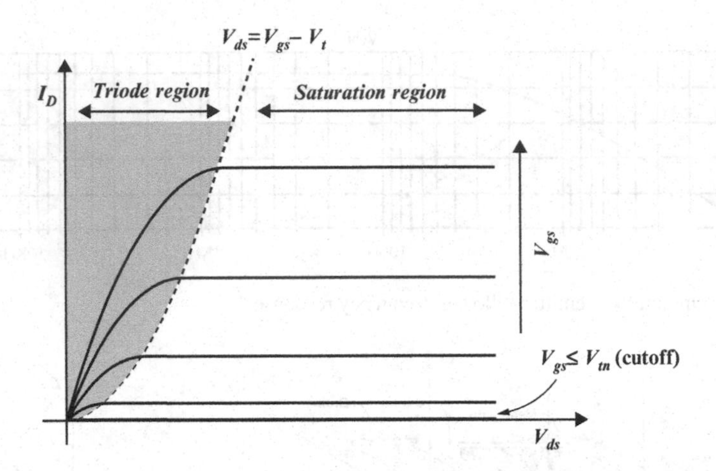

Fig. 1.15 Characteristic curve $I_d - V_{ds}$

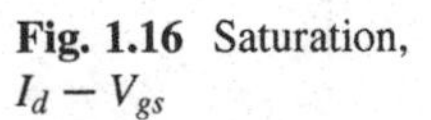

Fig. 1.16 Saturation, $I_d - V_{gs}$

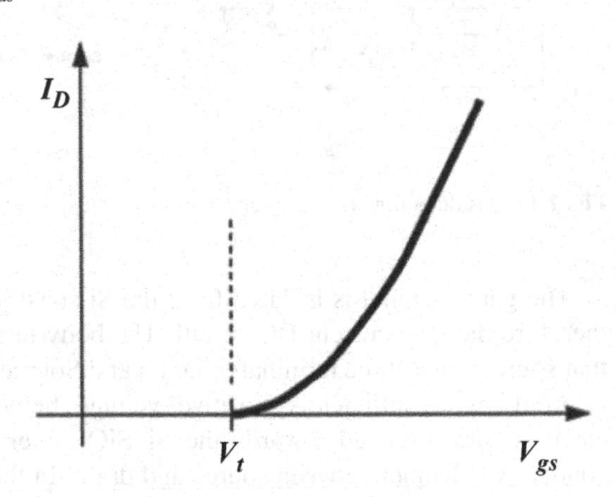

In most MOSFET applications, an input signal is the gate (G) voltage Vg and the output is the drain (D) current I_d. The ability of MOSFET to amplify the signal is given by the output/input ratio: the transconductance, $g_m = dI/dV_{gs}$.

In the Fig. 1.15 report the characteristic curve $I_d - V_{ds}$; there are three regions of work:

- Cutoff: in this case, it is necessary induce the channel, $V_{gs} \leq V_t$ (V_t is the threshold voltage) for nMOS.
- Triode: the channel must be induced and also keep Vds small enough so the channel is continuous (not pinched off): $V_{ds} \leq V_{gs} - V_t$.
- Saturation: in this mode need to induce the channel, $V_{ds} \geq V_{gs} - V_t$ then ensure that the channel is pinched off at the drain end. In Fig. 1.16 is visualized the plot of I_d versus V_{gs} for an enhancement type nMOS device in saturation.

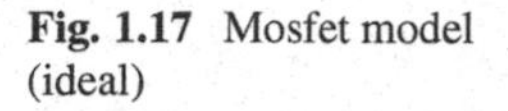

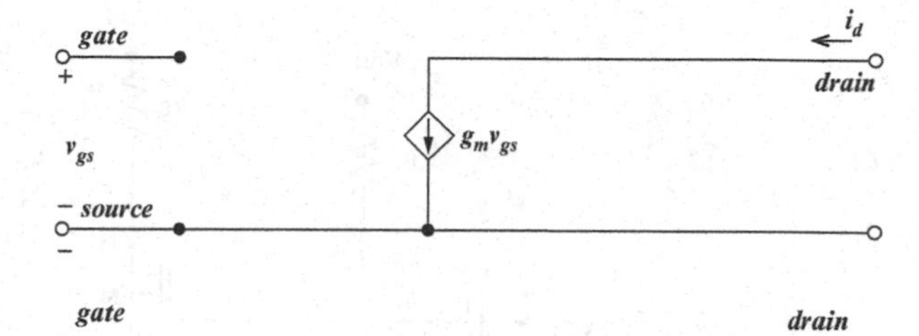

Fig. 1.17 Mosfet model (ideal)

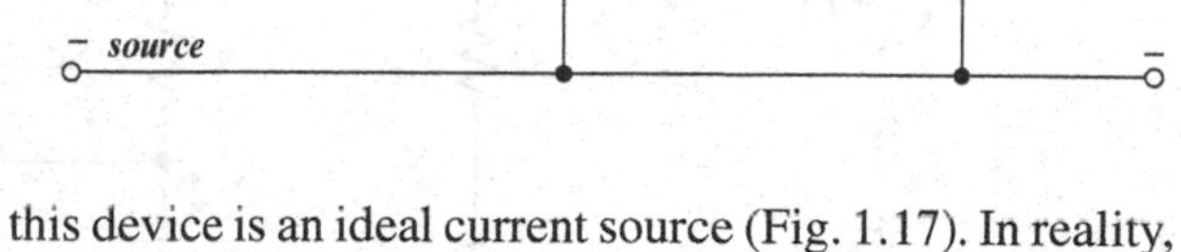

Fig. 1.18 Mosfet model (not ideal)

In the saturation mode, this device is an ideal current source (Fig. 1.17). In reality, there is to consider a finite output resistance (r_O); the outline of Fig. 1.18 can be described as in Fig. 1.17.

While the transconductance g_m gives the variations of the drain current due to variations of the voltage V_{gs}, there is another fundamental parameter of the Mosfet that takes the name of the output conductance, which is the variation of the drain current I_{ds} due to variations of the voltage V_{ds}.

The output conductance assumes a much greater importance, and the reason is the effect of the so-called channel length modulation due to V_{ds}: it is the effect for which the effective length of the channel decreases with increasing V_{ds} and in the saturation zone, the current increases with the V_{ds}. From an analytical point of view the effect of the modulation of the channel length can be written as follows: $I_{ds} = k(V_{gs} - V_t) * 2 * (1 + V_{ds}\lambda)$. It is clear that it provides a linear dependence of the I_{ds} vs V_{ds} in accordance with λ named as parameter of the channel length modulation; k is the transconductance factor proportional to the geometry of the Mosfet.

Example of application with Mosfet can be the common source amplifier visualized in Figs. 1.19 and 1.20. The input signal is applied to the gate through the coupling capacitor C_1. The output is on the drain and connected to the load through C_2, C_1, C_2, and C_S, which the coupling capacitors, and therefore can be considered short-circuits at the frequencies of the signal (center-band). The Source (S) is therefore to ground for the signals. R_1, R_2, R_D and R_S form a bias network to four resistors.

1.6 Differential Amplifiers

The main part of analog-integrated-circuit design is the differential-pair or differential amplifier configuration. The differential amplifiers were introduced in electronics to eliminate all or part of the problems of the amplifiers with direct

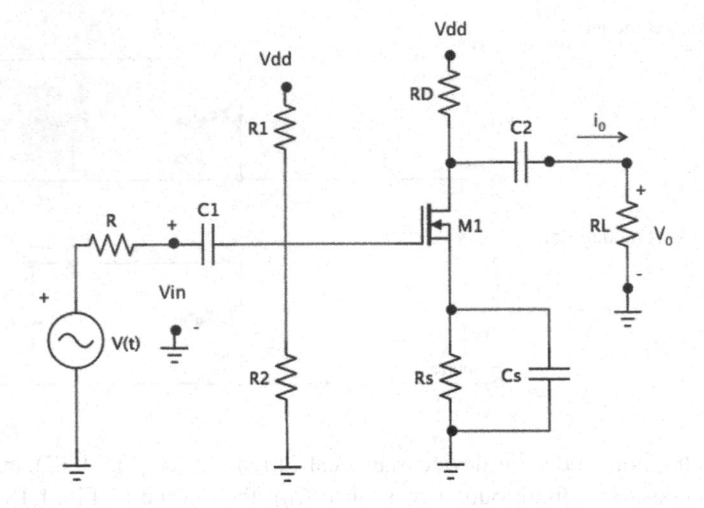

Fig. 1.19 Common source amplifier

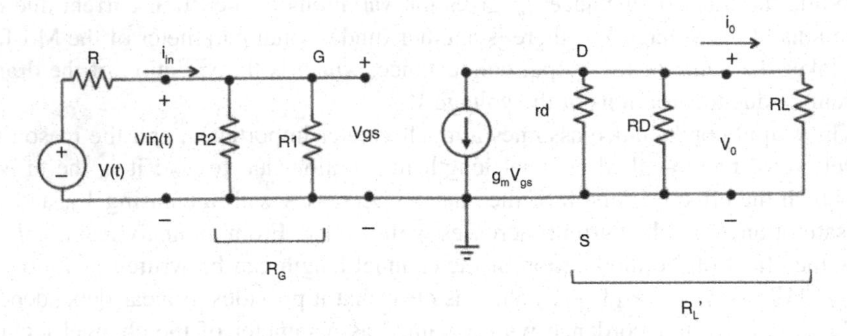

Fig. 1.20 Common source amplifier (model)

coupling. The goal is to create amplifiers characterized by an acceptable signal/noise ratio (SNR), also in the presence of external/internal disturbances generated by thermal variations due to the aging of electronics components. The differential-pair of differential-amplifier configuration is widely used in IC design. One example is the input stage of an op-amp and the emitter-coupled logic (ECL). This technology was invented in the 1940s for use in vacuum tubes; the basic differential-amplifier configuration was later implemented with discrete bipolar transistors. However, the configuration became most useful with the invention of the modern transistor/MOS technologies [3, 6–8].

The main features of differential amplifiers are as follows: (1) High input resistance, so small voltage signals can be amplified without losses. (2) Temperature drift is minimal. (3) Two input terminals, i.e. inverting and non-inverting inputs. (4) It amplifies the difference between two input signals.

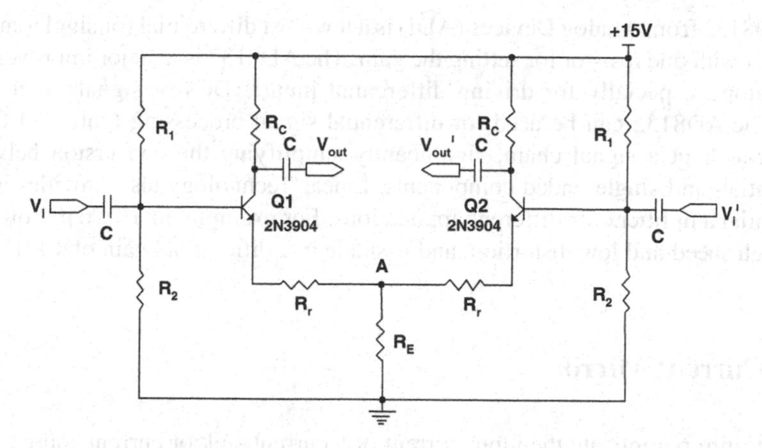

Fig. 1.21 Example of a differential amplifier

The differential amplifier works only on dual power supplies: it requires both $+V_{cc}$ and $-V_{cc}$ voltages simultaneously. However, if the same circuit is connected to a single power supply, its working becomes unstable and we do not get proper amplification from the circuit. The differential amplifier (Fig. 1.21) provides a number of advantages, making it one of the most useful circuit configurations, particularly as input stage for high gain and DC amplifiers. The figure of merit for differential amplifiers is called Common Mode Rejection Ratio (CMRR), defined as the ratio between difference mode gain (Ad) and common mode gain (Ac). Differential amplifiers is a special purpose amplifier designed to measure differential signals, otherwise known as a subtractor. The CMRR can be calculated with the following equation:

$$A_d = \frac{R_c}{2(R_r + r_{tr})}; A_c = \frac{R_c}{2R_E + R_r + r_{tr}}; \mathrm{CMRR} = \frac{2R_E + R_r + r_{tr}}{2(R_r + r_{tr})} \tag{1.18}$$

where r_{tr} is the trans-resistance, usually indicated also as r_e. The 741 (a common op-amp chip) has a CMRR of 90 dB, which is reasonable in most cases. A value of 70 dB may be adequate for applications insensitive to the effects on amplifier output; some high-end devices may use op-amps with a CMRR of 120 dB or more. The CMRR relates to the ability of the op-amp to reject a common-mode input voltage. This is very important because common-mode signals are frequently encountered in op-amp applications.

Common application of differential amplifiers is in the control of motors or servos, as well as in signal amplification. In discrete electronics, a common arrangement for implementing a differential amplifier is the long-tailed pair, typically found as a differential element in most op-amp integrated circuits. A long-tailed pair can be used as an analog multiplier with a differential voltage as one input and biasing current as another. There are many differential amplifier ICs on the market today.

The AD8132 from Analog Devices (ADI) is a low-cost differential (or single-ended) amplifier with one resistor for setting the gain. The AD8132 is a major improvement on op-amps, especially for driving differential input ADCs or signals over long lines. The AD8132 can be used for differential signal processing (gain and filtering) throughout a signal chain, significantly simplifying the conversion between differential- and single-ended components. Linear Technology also provides many differential amplifiers for different applications. For example, its LTC6409 offers a very high speed and low distortion, and is stable in a differential gain of 1 [7].

1.7 Current Mirror

Current mirrors replicate the input current of a current sink or current source in an output current, which may be identical or a scaled version. Current mirrors are used to provide bias currents and active loads to circuits [8]. Apart from some special circumstances, a current mirror (Figs. 1.22 and 1.23) is one of the basic building-blocks of the operational amplifier; it is a circuit designed to keep the output current constant regardless of loading. This type of topology may therefore be used in order to create current generators. The main requirements a current mirror must meet are:

- Output current independence of the output voltage.
- Wide range of output voltages at which the mirror is working properly.
- Low input voltage.

The range of voltages within which the mirror works is called the "compliance range," and the voltage marking the behavior in active/linear region is called the "compliance voltage." There are also a number of secondary performance issues with mirrors, such as temperature stability, for example.

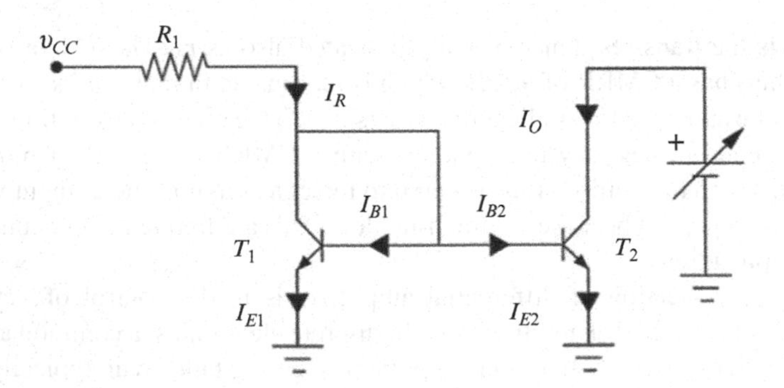

Fig. 1.22 Current mirror with BJT

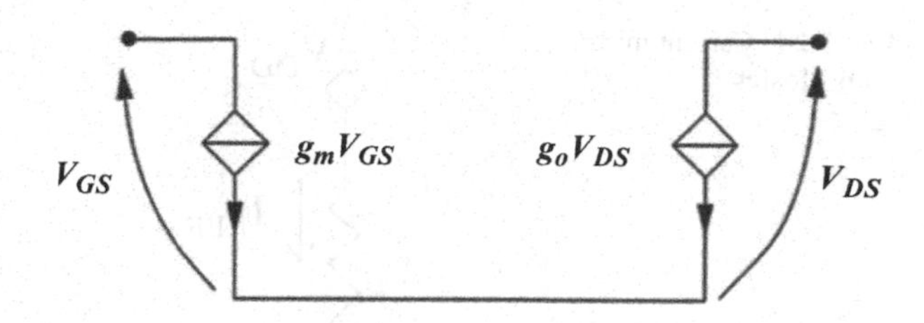

Fig. 1.23 Current mirror with Mosfet (model)

1.7.1 Ideal Current Mirror

A current mirror is usually and simply approximated by an ideal current source. However, an ideal current source is unrealistic for several reasons:

- It has an infinite AC impedance, whereas a practical mirror has a finite impedance;
- It provides the same current regardless of voltage, that is, there are no compliance range requirements;
- It has no frequency limitations, whilst a real mirror has limitations due to the parasitic capacitances of the transistors;
- The ideal source has no sensitivity to real-world effects like noise, power-supply voltage variations, and component tolerances;

The main part of the current mirror is a bipolar transistor (BJT) or a MOSFET. Transistors in a current mirror circuit must be maintained at the same temperature for precise operation. There are different types of current mirrors:

- Simple current mirror (BJT and MOSFET);
- Base current corrected simple current mirror;
- Widlar current source;
- Wilson current mirror (BJT and MOSFET);
- Cascode current mirror (BJT and MOSFET).

All of the circuits have compliance voltage that is the minimum output voltage required to maintain correct circuit operation: the BJT should be in the active/linear region and the MOSFET should be in the active/saturation region.

1.7.2 Current Mirror BJT/MOS

Current mirror circuits are usually designed with a BJT, such as an NPN transistor, where a positively doped (P-doped) semiconductor base is sandwiched between two negatively doped (N-doped) layers of silicon. These transistors are specifically designed to amplify or switch current flow. In some current mirror design specifications, the NPN transistor works as an inverting current amplifier, which reverses the current direction, or it can regulate a varying pulse current through amplification to

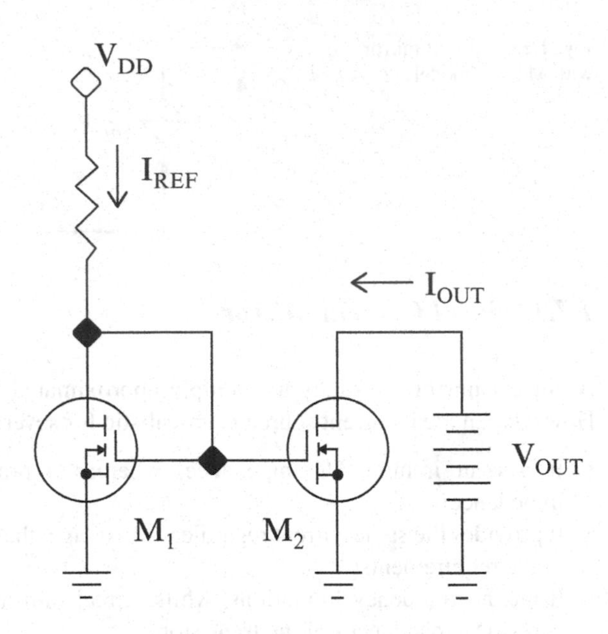

Fig. 1.24 Current mirror with Mosfet

create output mirror properties. One of the reasons that BJTs are used for current mirror design is due to the base–emitter (or PN part) of the transistor functioning reliably like a diode. Diodes regulate both the amount of current that passes and the forward voltage drop for that current. The basic current mirror can also be implemented using MOSFET transistors (Fig. 1.24). In Fig. 1.24, M_1 is operating in the saturation or active mode, and so is M_2. In this setup, the output current I_{OUT} is directly related to I_{REF}. The drain current of a MOSFET I_D is a function of both the gate-source voltage and the drain-to-gate voltage of the MOSFET given by a relationship derived from the functionality of the MOSFET. In the case of transistor M_1 of the mirror, $I_D = I_{REF}$. Reference current (I_{REF}) is a known current and can be provided by a resistor or by a "threshold-referenced" or "self-biased" current source to ensure that it is constant and independent of voltage supply variations [6–8].

References

1. Razavi B (2002) Design of analog CMOS integrated circuits. McGraw-Hill, New York
2. Di Paolo Emilio M (2013) Data Acquisition system from fundamentals to applied design. Springer, New York
3. Razavi B (2008) Fundamentals of microelectronics. Wiley, New York
4. Neudeck GW (1989) Modular series on solid state devices: volume III: the bipolar junction transistor, 2nd edn. Prentice Hall, Englewood Cliffs
5. Sedra AS, Smith KC (2013) Microelectronic circuits. Oxford University Press, New York
6. Razavi B (2002) Design of integrated circuits for optical communications. McGraw-Hill, New York
7. Hurst PJ (2001) Analysis and design of analog integrated circuits. Wiley, New York
8. Fonstad CG (1994) Microelectronic device and circuits. McGraw-Hill, New York

Chapter 2
Bipolar Transistor

Abstract This chapter will introduce the main element in the microelectronics, the transistor. The bipolar junction transistor is a device widely used in the analog electronics (and, sometimes, it has also been used in some digital electronics, although it will not be discussed here); it is an object typically planar, consists of a semiconductor doped: npn or pnp.

2.1 Ebers–Moll Model

The circuits on the modeling bipolar junction transistor (BJT) in static conditions are very complicated: they are highly nonlinear with exponential terms (equations of the diodes). The best way is to derive easier models, by introducing restrictions (Fig. 2.1). We suppose that the BJT is in the linear region of operation: in this way, a diode is ON (between the base and emitter), and another is OFF, described with an open circuit. A possible circuit can be one as represented in Fig. 2.2 by Ebers–Moll Circuit type T [5].

In Fig. 2.1 is shown the Ebers–Moll for NPN Transistor represented in Fig. 2.3. The two generators, V_{EB} and V_{CB}, provide bias voltages to the base–emitter and base–collector diodes. The currents I_E and I_C are described by the following two equations:

$$I_E = -I_{ES}(e^{V_{be}/\eta V_T - 1}) + \alpha_R I_{CS}(e^{V_{bc}/\eta V_T - 1}) \tag{2.1}$$

$$I_C = -I_{CS}(e^{V_{bc}/\eta V_T - 1}) + \alpha_F I_{ES}(e^{V_{be}/\eta V_T - 1}) \tag{2.2}$$

known as the Ebers–Moll equations where I_{ES} is the reverse saturation current of the base–emitter junction, the I_{CS} is the analogous to the base–collector junction. The signs are due to the convention adopted for the direction of the currents. The first term in the equation is due to the junction effect of the transistor; the second part, instead, would not appear in the equations if the system was only consists of two independent diodes.

The base–emitter junction is made with an asymmetrical doping, i.e., with a concentration of donors into the emitter much greater than that described acceptors in the base. Because of this, when the base–emitter junction is forward biased, the current is constituted for the almost totality of electrons that are injected from

M. Di Paolo Emilio, *Microelectronics*, DOI 10.1007/978-3-319-22545-6_2

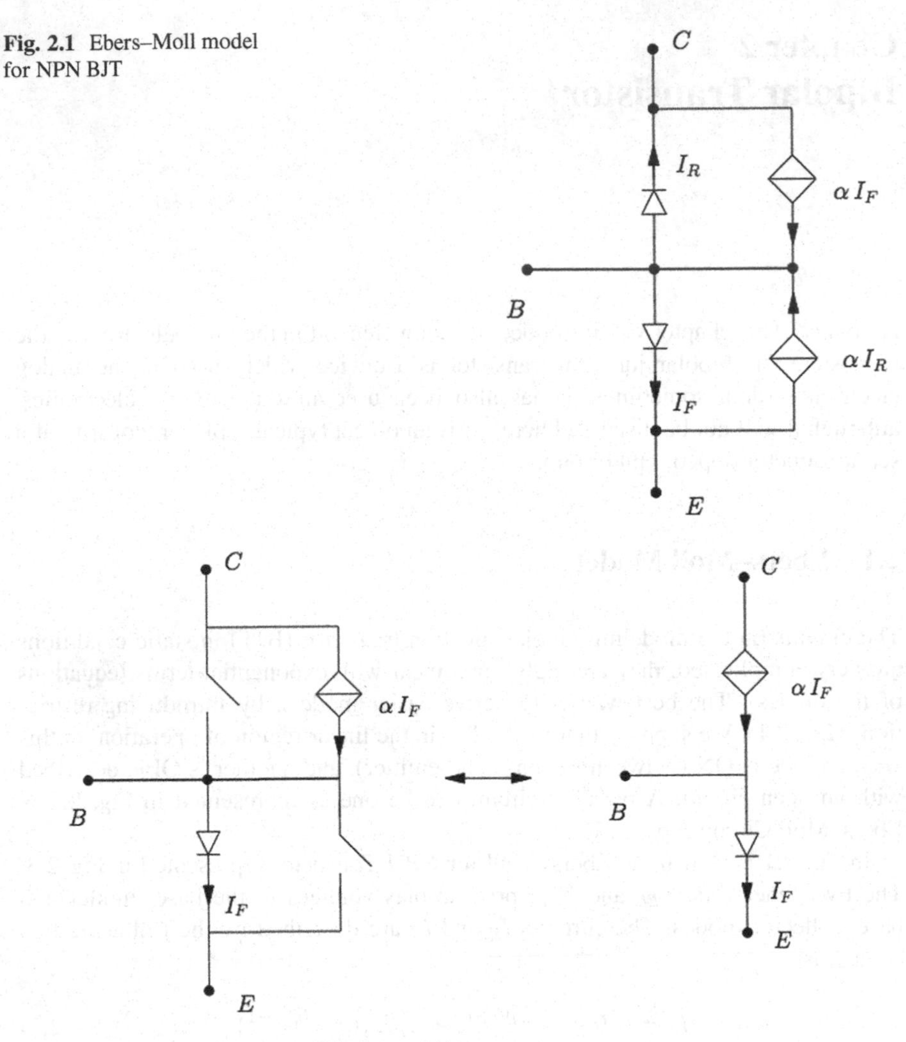

Fig. 2.1 Ebers–Moll model for NPN BJT

Fig. 2.2 Ebers–Moll model type T in forward bias

the emitter region in the base, while the contribution due to the shortcomings of the region of base to the emitter is negligible. Because of the thinness of the base region and of its geometry, a fraction of α_F of electrons injected into the base reaches the region of the collector by diffusion and contributes to the collector current I_C. The value of α_F is typically very close to 1, between 0.95 and 0.999. Similarly, when it is forward biased junction collector–base, a fraction α_R of carriers are injected from the collector in the base and reach the emitter and give a contribute to the current I_E; the value of α_R is generally quite small, typically of the order of 0.5. This asymmetry between α_F and α_R is due to the structural features of the two junctions [3, 5, 7].

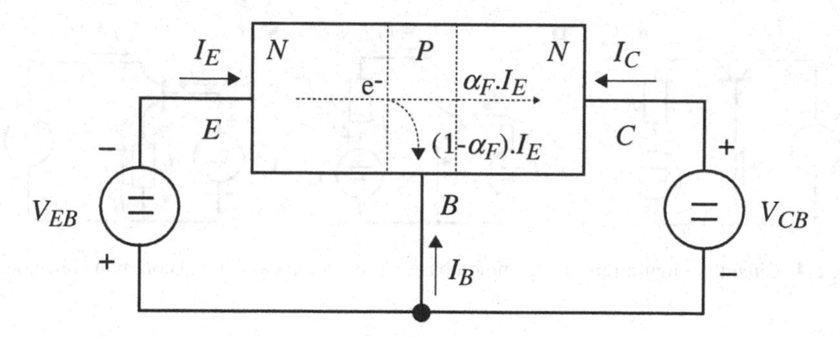

Fig. 2.3 NPN transistor

2.2 Operations Mode

The BJT is said off (cutoff) when both junctions are inversely biased and therefore the only circulating currents are the two weak reverse saturation current I_{ES} and I_{CS} of the two junctions. A transistor BJT works in the active region when the base–emitter junction is forward biased and the base–collector is inversely biased. In these conditions, the first term of Eq. (2.1) and the second of (2.2) are predominant and the two equations can be simplified in:

$$I_E = -I_{ES}(e^{V_{be}/\eta V_T}) \tag{2.3}$$

$$I_C = \alpha_F I_{ES}(e^{V_{be}/\eta V_T}) = -\alpha_F I_E \tag{2.4}$$

Using the Kirchhoff law, we can obtain the base current:

$$I_B = -I_E - I_C = I_C \frac{1-\alpha_F}{\alpha_F} \tag{2.5}$$

Then:

$$I_C = \beta_F I_B \tag{2.6}$$

$$I_E = -(\beta_F + 1)I_B \tag{2.7}$$

with:

$$\beta_F = \frac{\alpha_F}{1-\alpha_F} \tag{2.8}$$

α_F has values very close to one, the value of β_F is much greater than one, typically between 20 and 1000. Dropping to zero the bias voltage V_{CB} and subsequently reversing the polarity, the collector–base junction will be switched from interdiction to conduction, for which the first term of Eq. (2.2) will not be more significant. Because of the exponential dependence of the currents from the bias voltages, for a value of V_{CB} very close to that of V_{EB}, i.e. for $V_{CE} \sim 0$, the two terms

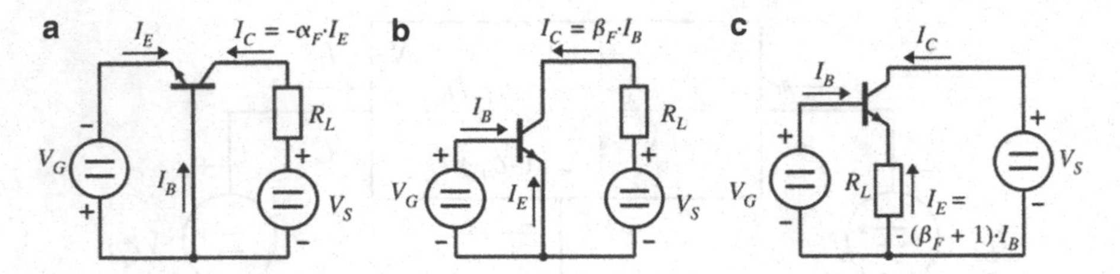

Fig. 2.4 Circuit configurations: (**a**) common base, (**b**) common emitter, (**c**) common collector

have the same value giving a total $I_C = 0$. A BJT used as an amplifier is normally biased to work in the active region. The two regions of interdiction and saturation regions are typically used as a switch in digital circuits [4].

2.3 Circuit Configurations

The transistor is a device with three terminals. Therefore, in an amplifier configuration, one of the three terminals will be in common to both the ports, the input and output,which is connected to the ground. There are three different circuit configurations: common base (common base or mass), common emitter, and common collector (Fig. 2.4).

2.3.1 *Common Base*

The input terminal is the emitter and the output is the collector (Fig. 2.4a). The input current I_E, sent by the generator V_G, is transferred from the transistor in the output circuit as a current I_C, with a current amplifier $-\alpha_F$ slightly less than one and does not depend on the voltage V_{CB}. The voltage is generated across the resistance of the load R_L, if the value of R_L is chosen correctly, is much larger than the input voltage to the emitter. This has historically been the first configuration of use of BJT.

2.3.2 *Common Emitter*

The input terminal is the base and the output is the collector (Fig. 2.4b). The generator V_G polarizes the base–emitter junction exactly as in the case of the configuration with the base to ground, but this time, it delivers only the weak base current I_B, which is sometimes smaller current I_C. A BJT with grounded emitter is a current amplifier with an amplification factor β_F. For these reasons, the configuration with common emitter is to use more frequently.

2.3.3 Common Collector

The input terminal is the base and the emitter is the output (Fig. 2.4c). The current flowing in the load resistance R_L is $\beta_F + 1$ times that of the base: the transistor with collector to ground is a current amplifier. The voltage amplification, instead, is slightly less than one, since the voltage in output on the load R_L, is the generator input V_G minus the fraction required to polarize the base–emitter junction. The circuit with common collector is also called voltage follower (voltage follower or emitter follower).

2.4 Characteristic Curves

In Fig. 2.5 is shown in graphical form the relationships between currents and voltages of the input and output for a BJT NPN in common emitter configuration, as provided by equations Ebers–Moll. The curves in Fig. 2.5a describe the collector current I_C as a function of collector–emitter voltage V_C and of the base current I_B. The relation $I_C(I_B)$ with V_{CE} a constant is the characteristic of direct transfer; instead, the relation $I_C(V_{CE})$ with I_B a constant is the output characteristic.

In Fig. 2.5b is shown the relationship between the base current I_B and base–emitter voltage V_{BE}. This graph expresses two reports: the input characteristic $I_B(V_{BE})$ and the transfer characteristic inverse $I_B(V_{CE})$. Moreover, the graph contains only one curve because, according to the equations of Ebers–Moll, the relation $I_B(V_{BE})$ is practically independent from V_{CE} across the active region [8].

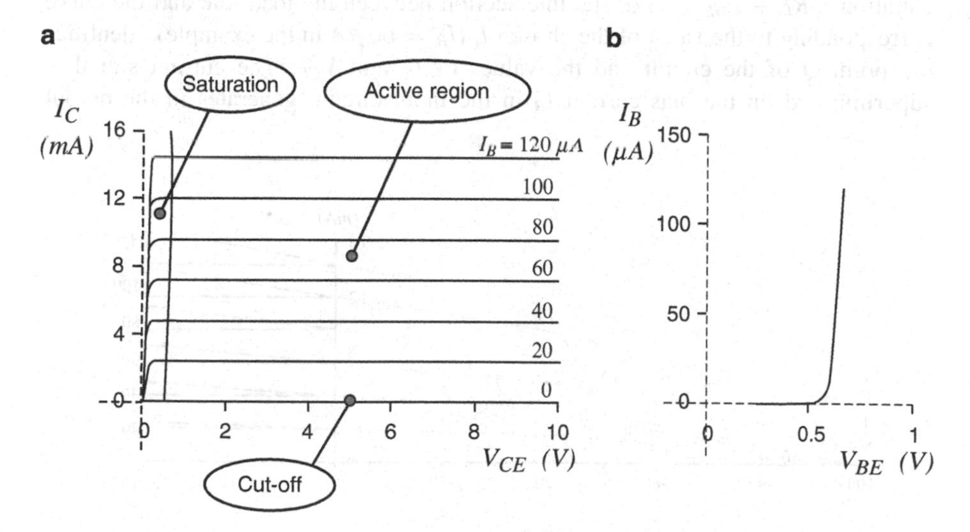

Fig. 2.5 Characteristics curves: (**a**) output, (**b**) input

2.5 Early Effect

Ebers–Moll equations predict that in the active region current I_C depends only on the voltage V_{BE} or the current I_B, and this is evident in the graph in Fig. 2.5a. A more description of the behavior of a transistor BJT is shown in Fig. 2.6: in the active region, the current I_C has an approximately linear dependence from V_{CE}, described by the equation:

$$I_C = \beta_F I_B \left(1 + \frac{V_{CE}}{V_A}\right) \tag{2.9}$$

This dependence is due to the Early effect: the thickness of the transition region of the base–collector junction varies with the polarization and this causes a change in the parameter α_F (and therefore β_F). The voltage $-V_A$ corresponds to the point where they meet the straight lines obtained by the extension to the left of the characteristic curves in the active region and is said Early voltage.

2.6 Common Emitter Amplifier

The common emitter configuration is shown in the diagram of the Fig. 2.7a and is the most used at the same time as amplifier of voltage and current. The two current generators are connected to the base circuits that provide the bias current I_B to determine the working point. The operation is illustrated graphically in Fig. 2.7b, which shows the characteristic curves $I_C(I_B, V_{CE})$ and the load of line equation $I_C R_L + V_{CE} = V_S$. The intersection between the load line and the curve corresponding to the value of the chosen $I_B (I_B = 60\,\mu A$ in the example), identifies the point Q of the circuit and the values of I_C and V_{CE}. The current signal i_b superimposed on the bias current I_B in the input circuit, generates in the output

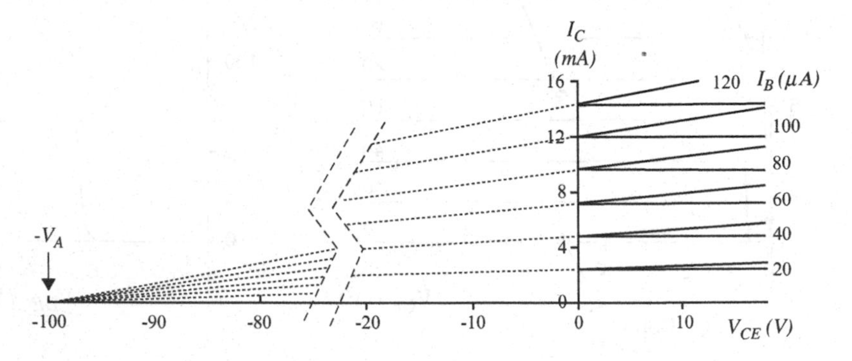

Fig. 2.6 Ebers–Moll equation in comparison with Early effect

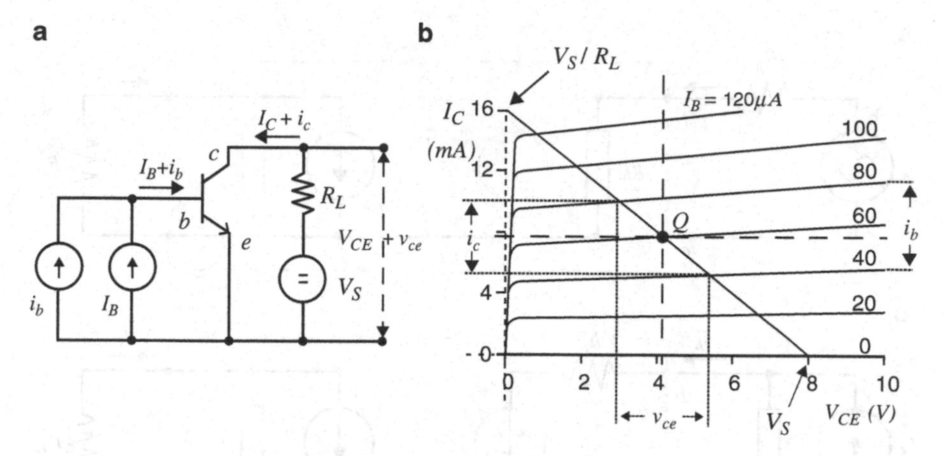

Fig. 2.7 Common emitter amplifier: (**a**) layout, (**b**) characteristics curve

circuit the variations i_c and v_{ce} indicated in the Fig. 2.7b. The input voltage v_{be}, corresponding to the current i_b, It may be determined by the curve of Fig. 2.7b. The direct use by numerical calculation (or in graphical form) of the equations of Ebers–Moll is only convenient in the presence of signal excursions large to bring the circuit in the non-linearity region, for example in the study of the power circuits [6].

Once the work point has determined the values of the polarization currents and voltages, the circuit of Fig. 2.7 may be linearized and then re-designed to contain only items related to the signals. The equations of Ebers–Moll suggest to choose the collector current I_C as the output dependent variable and consequently the voltage V_{CE} as the independent variable. As regards the input is equally convenient to choose as the independent variable both the voltage (and therefore the parameters g) and the current. The related circuits are shown in Fig. 2.8. Each parameter of model is added to the index and specified that it refers to a transistor used with common emitter. Depending on the model chosen, the signal source is represented by an equivalent generator of Thevenin, with an internal resistance in series (v_g, R_g), or by an equivalent generator of Norton, with a conductance internal in parallel (i_g, G_g).

The two most important elements of the linear model are the input parameter ($g_{i.e.}, h_{i.e.}$) and the parameter of direct transfer(g_{fe}, h_{fe}); the other two elements can be considered zero in the first approximation. The fact that the output conductance (g_{oe}, h_{oe}) is zero in the first approximation clarifies why the parameters g and h are the most convenient in the description of BJT: the models related to the parameters m and r provided in the output circuit a voltage generator. So, in the first approximation, we would have to insert a voltage source in series with an infinite resistance in order to simulate the behavior of the collector circuit, which is a current generator [5, 6].

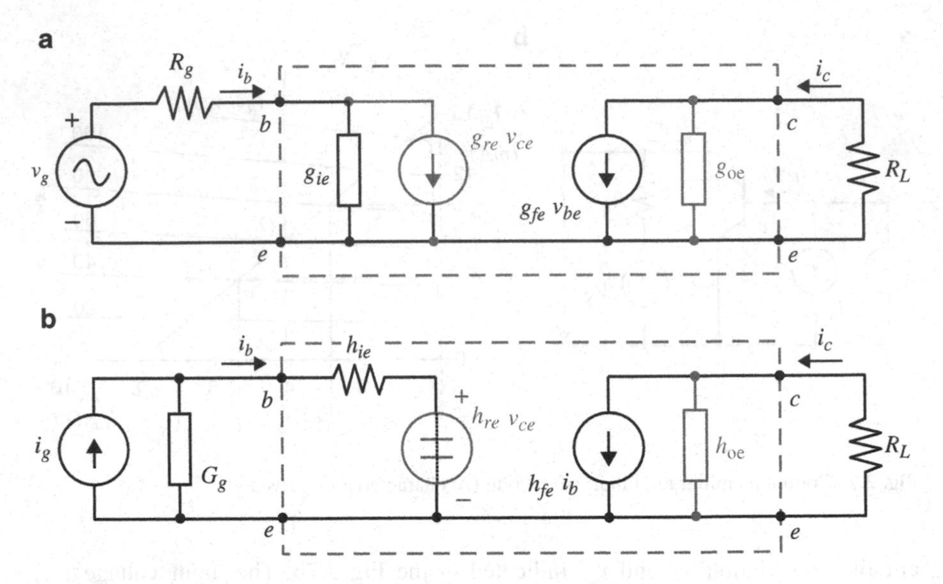

Fig. 2.8 Linear model with g parameters (**a**) and h parameters (**b**)

In the circuits of Fig. 2.8 the input signal generators and the output load resistance R_L are not properly a part of the amplifier, but represent the external circuits to which the amplifier is connected. The generator v_g thus represents the amplitude of the input signal to empty, i.e.: when the amplifier is not yet connected to the source. Since, the amplifier presents an input conductance $g_{i.e.}$ that has an effect of partition, with a reduction of the signal voltage by a factor $1/(1 + g_{i.e.}R_g)$. Similarly, for the current generator i_g it has an effect of partition between G_g and $h_{i.e.}$ with a reduction of the input current by a factor $1/(1 + h_{i.e.}G_g)$. Taking into account of these effects, the voltage gain for the circuit becomes:

$$A_v^{'} = \frac{-g_{fe}R_L}{(1 + g_{oe}R_L)(1 + g_{ie}R_g)} \tag{2.10}$$

while the current gain remains unchanged. The current gain for the circuit Fig. 2.8 is the following equation:

$$A_i^{'} = \frac{-h_{fe}}{(1 + h_{oe}R_L)(1 + h_{ie}G_g)} \tag{2.11}$$

Please note that the two quantities are not intrinsic properties of the amplifier, but they describe the behavior in its interaction with the external circuits. The correct way to specify the characteristics of the amplifier is to give the values of $A_i$and A_v together with the input resistance. The records shall contain all the information necessary to know the behavior of the circuit in its interaction with a signal source.

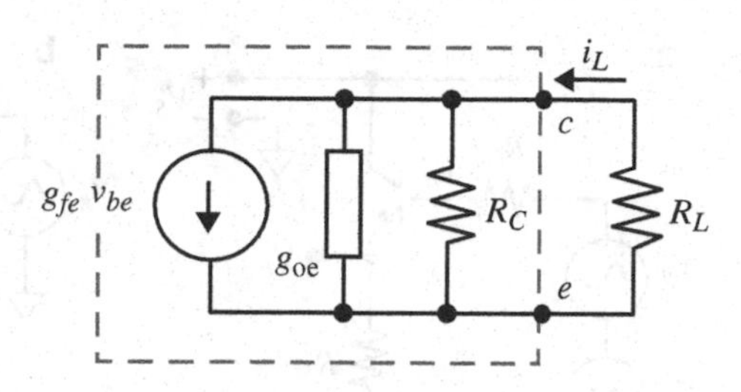

Fig. 2.9 Output circuit with load resistance R_L and collector resistance R_C

A similar problem occurs in the output circuit. Up to now the resistance R_L was considered an integral part of the amplifier circuit itself. In fact, by examining the bias circuits then it will see that the collector can be presented as resistance R_C, that it is not the actual user of the signal, but is required for the polarization. The load resistance itself is instead a further resistor R_L connected to the output (Fig. 2.9). Even in this case, it will have an effect of partition of the current between the output resistance r_o of the amplifier itself, which is given by the parallel of R_C with the output conductance g_{oe}, and the load resistance R_L:

$$A_v'' = \frac{A_v R_L}{R_L + r_o} \tag{2.12}$$

$$r_o = \frac{R_c R_C}{1 + g_{oe} R_C} \tag{2.13}$$

2.7 Common Collector Amplifier

In the circuit in Fig. 2.10, the load resistance R_L has been inserted on the emitter and the collector, which is connected to ground with respect to the signals. In fact, in the linearized model for small signals (Fig. 2.10b), the generator V_S is replaced by its internal resistance, i.e. by a short circuit to ground. The linearized circuit of Fig. 2.10b contains only the two main terms $h_{i.e.}$ and h_{fe}:

$$i_e = -(1 + h_{fe}) i_b \tag{2.14}$$

for which the potential of the emitter is:

$$v_e = (1 + h_{fe}) R_L i_b \tag{2.15}$$

and that of the base is:

$$v_b = h_{ie} i_b + (1 + h_{fe}) R_L i_b \tag{2.16}$$

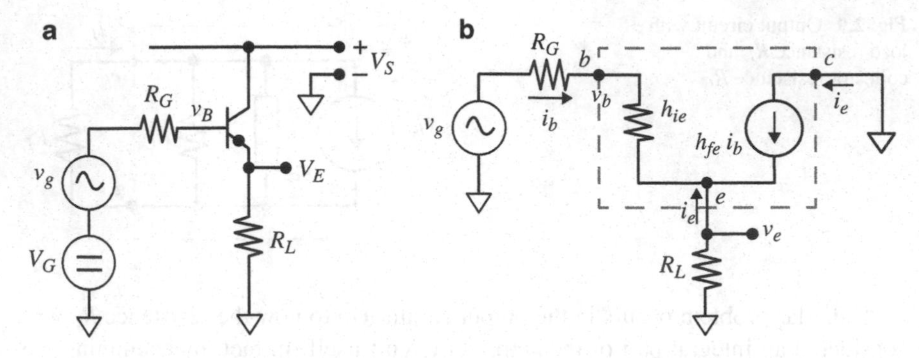

Fig. 2.10 Emitter follower: (**a**) general layout (**b**) Linearized model for small signals

Compared to the generator v_g, we can have the input resistance of h_{ie} in series with $(1 + h_{fe})R_L$. Then, the emitter voltage is the following equation:

$$v_e = v_b * \frac{(1 + h_{fe})R_L}{h_{ie} + (1 + h_{fe})R_L} \tag{2.17}$$

where usually R_L and $h_{i.e.}$ are of the same order of magnitude while h_{fe} is of the order of 100, $v_e \sim v_b$ with a difference of the order of percent (or a few percent) and v_e slightly less than v_b. This is the reason for the name given to this circuit: emitter follower. The voltage outputs on emitter closely follow the input voltage, remaining always just below.

Assuming to put $v_g = 0$ and apply from outside the tension on the emitter, the current will be the following equation:

$$i_e = -i_b * (1 + h_{fe}) = \frac{v_e(h_{fe} + 1)}{h_{ie} + R_G} \tag{2.18}$$

which gives the following output resistance r_o:

$$r_o = \frac{v_e}{i_e} = \frac{h_{ie} + R_G}{h_{fe} + 1} \tag{2.19}$$

The behavior of the emitter follower is described by the following equations:

$$A_v = \frac{v_{ec}}{v_{bc}} = \frac{(1 + h_{fe})R_L}{h_{ie} + (1 + h_{fe})R_L} \sim 1 \tag{2.20}$$

$$A_i = \frac{i_e}{i_b} = -(h_{fe} + 1) \tag{2.21}$$

$$r_i = \frac{v_{bc}}{i_b} = h_{ie} + (1 + h_{fe})R_L \tag{2.22}$$

$$r_o = \frac{v_{ec}}{i_e} = \frac{h_{ie} + R_G}{h_{fe} + 1} \tag{2.23}$$

For these characteristics, the emitter follower is widely used as a current amplifier and impedance adapter [1, 2].

2.8 Common Base Amplifier

Figure 2.11 shows the basic common-base circuit, in which the base is at ground and the input signal is applied to the emitter.

Figure 2.12a again shows the hybrid-model of the npn transistor, with the output resistance infinite. Figure 2.12b shows the small-signal equivalent circuit of the common-base circuit, including the hybrid-model of the transistor.

As a result of the common-base configuration, the hybrid-model in the small-signal equivalent circuit may look a little bit of strange.

The small signal output voltage is given by:

$$V_o = -(g_m V_\pi (R_C R_L) \tag{2.24}$$

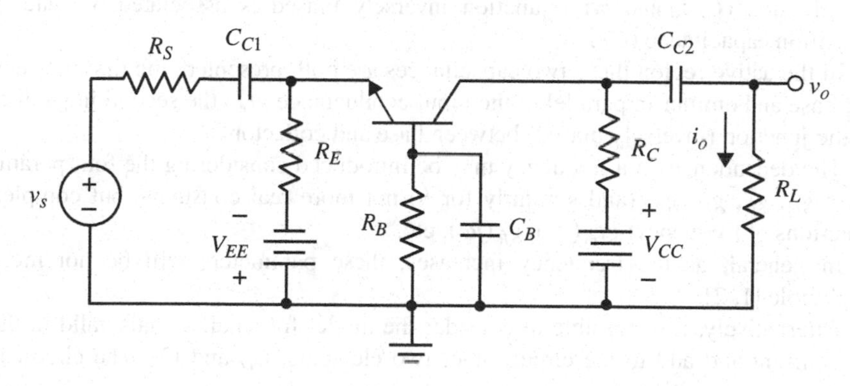

Fig. 2.11 Common base amplifier

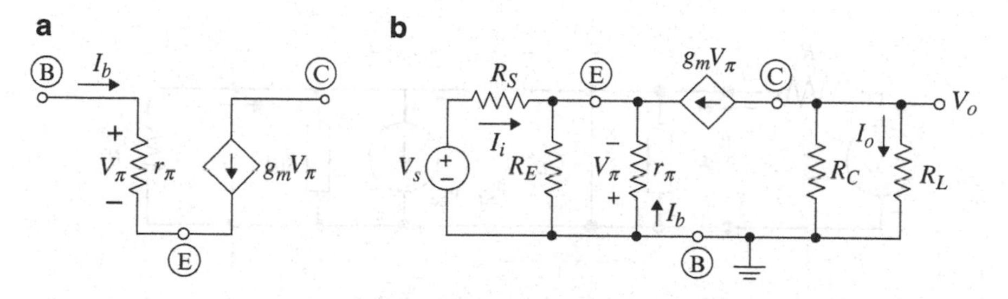

Fig. 2.12 (**a**) Simplified hybrid-π mode and (**b**) small-signal equivalent circuit of the common base circuit

and small-signal voltage gain, as follows:

$$A_v = \frac{V_o}{V_s} = g_m \frac{R_C R_L}{R_S} \left(\frac{r_\pi}{1+\beta} || R_E || R_S \right) \tag{2.25}$$

Figure 2.12b can also be used to determine the small-signal current gain. It is given by the following equation:

$$A_v = \frac{I_o}{I_i} = g_m \frac{R_C}{R_C + R_L} \left(\frac{r_\pi}{1+\beta} || R_E \right) \tag{2.26}$$

2.9 BJT in High Frequency

Linear models for the two-port devices described above are mathematical abstractions used to describe the behavior of the BJT in the region of small signals. In these models it is not contained no explicit dependence of the behavior of the transistor from the signal frequency.

On the other hand, with a forward biased PN junction is associated a capacitance of diffusion (C_D), and with junction inversely biased is associated, instead, a transition capacitance (C_T).

In the active region these two capacitances are both presented, the first between the base and emitter in parallel to the input conductance $g_{i.e.}$, the second in parallel to the junction (inversely biased) between base and collector.

The dependence on a frequency may be introduced considering the four parameters g_{ie}, g_{fe}, g_{re}, g_{oe} (and similarly for h) not more real constants, but complex functions of frequency: $g_{i.e.}(j\omega)$, $g_{fe}(j\omega)$, etc.

In general, as the frequency increases, these parameters will be not more negligible [1, 2].

Alternatively, it is possible to consider the model for small signals valid in the DC current and add to the circuit other two elements, C_D and C_T. The circuit is visualized in Fig. 2.13.

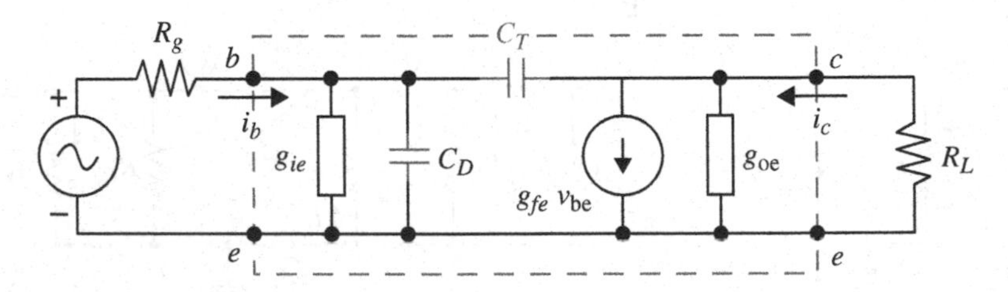

Fig. 2.13 Linear model of the BJT in the active region with diffusion capacitance of base–emitter junction and transition capacitance C_T of base–collector junction

2.9.1 Transition Frequency

In the circuit of Fig. 2.13 the collector is a circuit short with the emitter, i.e. placing $R_L = 0$, the voltage gain A_v is reduced to zero and the capacitance C_T is in parallel to C_D. In these conditions, the current gain is the following equation:

$$A_i = \frac{i_c}{i_b} = \frac{g_{fe}}{g_{ie} + j\omega(C_D + C_T)} \tag{2.27}$$

In DC current and in very low frequency, the current gain is constant and equal to $g_{fe}/g_{ie}{=}h_{fe}$; as the frequency increases, there is a trend of low-pass type, having a cutoff frequency:

$$f_\beta = \frac{g_{ie}}{2\pi(C_D + C_T)} \tag{2.28}$$

For frequencies much higher of f_β, the current gain can be written in the following way:

$$|A_i| = \frac{g_{fe}}{\omega(C_D + C_T)} \tag{2.29}$$

The frequency f_T where $|A_i| = 1$ is called the transition frequency:

$$f_T = \frac{g_{fe}}{2\pi(C_D + C_T)} \tag{2.30}$$

The relationship between f_T and f_β is the following:

$$\frac{f_T}{f_\beta} = \frac{g_{fe}}{g_{ie}} = h_{fe} \tag{2.31}$$

The frequency f_T and the coefficient of current amplification h_{fe} are two fundamental parameters for a BJT and they are normally specified by the manufacturer.

2.9.2 Miller Effect

A complete calculation of the behavior of the circuit of Fig. 2.13 can be done by writing and solving the linear system of equations (complex) of the circuit. An approximate calculation, but sufficient to identify the role of the various elements and evaluate time constants and cutoff frequencies, can be done by modifying the circuit as in Fig. 2.14.

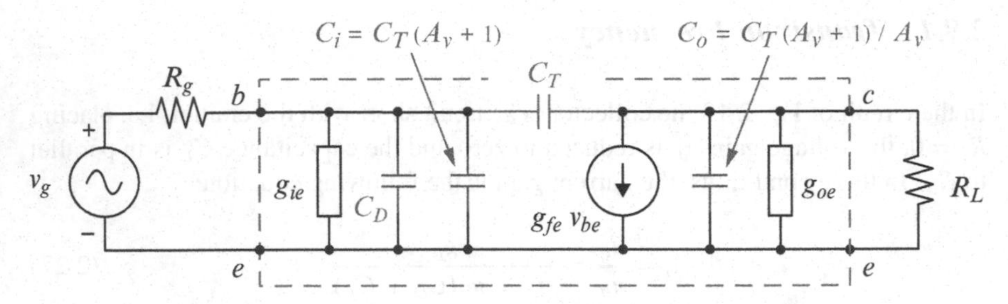

Fig. 2.14 Linear model of the BJT in the active region with transition capacitance C_T of base–collector junction replaced with equivalent capacitances C_i and C_o in accordance with the Miller effect

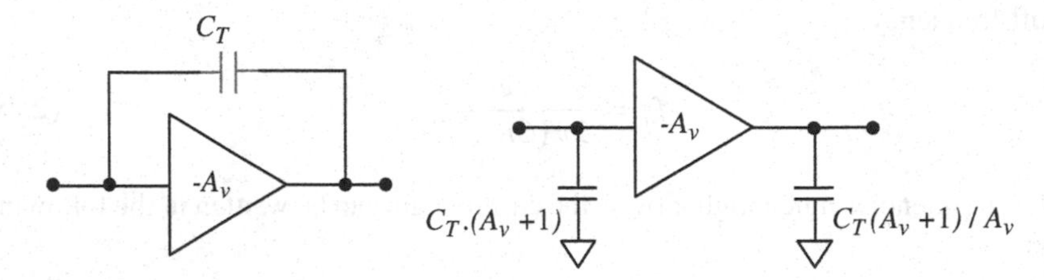

Fig. 2.15 Miller effect

The capacitance C_T is located between the output and the input of amplifier with voltage gain A_V. Assuming that the output of the amplifier is like a good voltage source, that is equivalent to have an impedance small compared to the reactance of the capacitor C_T, the circuit behaves as if there is a capacitance $C_i = C_T * (1 + A_V)$ between the input of the amplifier and ground, and a capacitance $C_o = C_T * A_V/(1 + A_V)$ between the output and ground (Fig. 2.15). The input capacitance is in parallel and will be added to C_D to reduce the high cutoff frequency of the amplifier [1, 2].

2.9.3 Bandwidth of an Amplifier

Taking into account for example: a common emitter amplifier with a transistor with characteristics and working point as shown in table of Fig. 2.16. The transition capacitance C_T is much more smaller than C_D (4.5 pF versus 65 pF) and its contribution to the input capacitance is largely predominant (C_i = 1.1 nF) due to the Miller effect. The constant time of the input circuit is:

$$\tau_i = \frac{(C_D + C_i)}{g_{ie}} = 0.54\,\mu\text{s} \tag{2.32}$$

Common emitter Amplifier with BJT			
Characteristics of transistor	h_{fe}	120	
	f_T	550 MHz	
	C_T	4.5 pF	
	g_{oe}	70 $\mu A/V$	
Working point	I_E	7 mA	
	R_L	1 $k\Omega$	
	R_g	50 Ω	
Parameters	g_{fe}	240 mA/V	$g_{fe} = I_{E/\eta}V_T$
	g_{ie}	2 mA/V	$g_{ie} = g_{fe}/h_{fe}$
	A_V	224	$A_V = g_{fe}/(g_{oe} + R_L^{-1})$
	C_D	65 pF	$C_D = g_{fe}/2\pi f_T - C_T$
	C_i	1.01 nF	$C_i = C_T \cdot (1 + A_V)$
	C_o	4.5 pF	$C_o = C_T \cdot (1 + A_V)/A_V$

Fig. 2.16 Parameters of emitter common amplifier

And the constant time of the output circuit is given by the following equation:

$$\tau_o = \frac{C_o}{g_{oe} + R_L^{-1}} = 4.2\,\text{ns} \tag{2.33}$$

The two cutoff frequency corresponding are the following:

$$f_{Hi} = \frac{1}{2\pi\tau_i} = 290\,\text{kHz} \tag{2.34}$$

$$f_{Ho} = \frac{1}{2\pi\tau_o} = 38\,\text{MHz} \tag{2.35}$$

The main factor in the response of high frequency is due to the effect Miller capacitance. In the input circuit, the internal resistance R_g of the generator is not considered. This is equivalent to assume a signal generator as a current generator with internal resistance $R_g = \infty$.

The graph of Fig. 2.17 shows the amplification calculated as the ratio between output voltage and input current (r_{21}). The bandwidth is limited to 290 kHz, as expected (Fig. 2.18).

The bandwidth can be extended up to 3.5 MHz using a signal generator with input impedance of 50 ω. This is explained by the fact that the time constant is passed from 580 to 53 ns due to the input impedance of the generator in parallel with the $g_{i.e.}$, bringing the cutoff frequency F_{HI} to 3.0 MHz.

Apparently the bandwidth of the circuit is increased. In practice, the generator is driving the base of the transistor in voltage, by supplying more current in proportion to the frequency [1, 2].

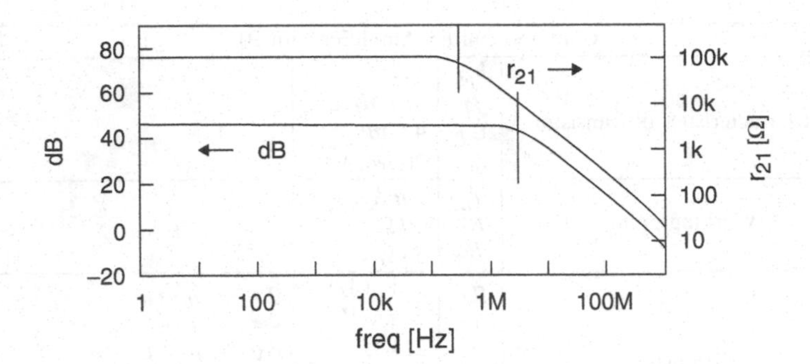

Fig. 2.17 Bandwidth for common emitter amplifier

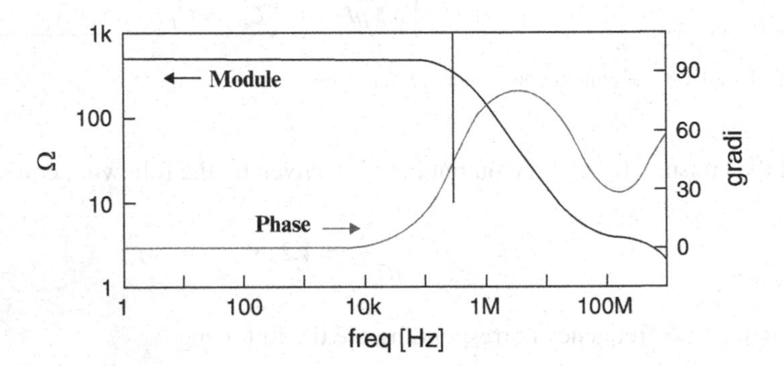

Fig. 2.18 Module and phase of the input impedance as function of the frequency for common emitter amplifier

References

1. Fonstad CG (1994) Microelectronic device and circuits. McGraw-Hill, New York
2. Hurst PJ (2001) Analysis and design of analog integrated circuits. Wiley, New York
3. Di Paolo Emilio M (2013) Data Acquisition system from fundamentals to applied design. Springer, New York
4. Neudeck GW (1989) The bipolar junction transistor. Modular series on solid state devices, vol. III, 2nd edn. Prentice Hall, Englewood Cliffs
5. Razavi B (2002) Design of analog CMOS integrated circuits. McGraw-Hill, New York
6. Razavi B (2002) Design of integrated circuits for optical communications. McGraw-Hill, New York
7. Razavi B (2008) Fundamentals of microelectronics. Wiley, New York
8. Sedra AS, Smith KC (2013) Microelectronic circuits. Oxford University Press, New York

Chapter 3
MOSFET

Abstract The MOSFETs, acronym for Metal-Oxide-Semiconductor Field-Effect-Transistor (transistor field-effect-type metal-oxide-semiconductor), owe their ever increasing popularity to the excellent electrical characteristics, such as high input impedance. The main applications are in the automotive, computer peripherals, motor control, and Switched Mode Power Supplies. The power MOSFETs on the market today perform the same function of the bipolar transistors (BJT), with more advantages and a technique of operation based on the voltage control. The continuous research has improved its characteristics for the complete replacement in the field of BJT.

3.1 Introduction

A field effect transistor (FET) works as a conduction channel in the semiconductor with two ohmic contacts, drain and source, where the number of charge carriers in the channel is controlled by a third contact, the gate. The most important FET is the MOSFET. In a MOSFET of silicon, the gate is separated from the channel by an insulating layer of silicon dioxide (SiO_2). The charge carriers of the conductive channel constitute a charge inversion, i.e. electrons in the case of a p-type substrate (n-channel device) or holes in the case of an n-type substrate (p-channel device), that are induced in silicon-insulator interface by the voltage applied to the gate electrode. MOSFETs are used both as discrete components that active elements in digital and analog monolithic integrated circuits (IC) [1].

3.2 General Characteristics of a MOSFET

The structure of a traditional metal-oxide-semiconductor (MOS) is obtained by growing a layer of silicon dioxide (SiO_2) on top of a silicon substrate, depositing in turn a layer of metal or polycrystalline silicon. Since the silicon dioxide is a dielectric material, its structure is equivalent to a planar capacitor, with one of the electrodes replaced by a semiconductor. A transistor of metal-oxide-semiconductor field effect (MOSFET) is based on the modulation of charge concentration by MOS capacitor between an electrode and the gate located over the substrate (body) and

M. Di Paolo Emilio, *Microelectronics*, DOI 10.1007/978-3-319-22545-6_3

isolated from all other regions of the device to a dielectric layer which, in the case of a MOSFET, is an oxide. Compared to the MOS capacitor, the MOSFET includes two additional terminals (source and drain), each is connected to the single highly doped regions and separated from the body region or substrate (substrate). Such regions may be of the type p or n. If the MOSFET is an n-channel or nMOSFET, then the source and drain are "n +" and the substrate is a region "p". If the MOSFET is a p-channel or pMOSFET, then the source and drain regions are "p +" and the substrate is a region "n". By applying a sufficiently positive voltage between gate and source, the electrons are attracted towards the Si–SiO_2 interface under the gate, forming a conductive channel between source (S) and drain (D). In these conditions, if a voltage is applied $V_{ds} > 0$, a current will flow between drain and source controlled by the voltage V_{gs} between gate and source, that controls the formation of the channel. The electrical characteristics of the MOSFET depend on L (gate length) and W (gate width), as well as the technological parameters such as oxide thickness and doping of the body. Typical values of L and W are: $L = 0.1$–2 pm, $W = 0.5$–500 μm. The range of the gate oxide thickness is 3–50 nm. There are four types of MOS transistors: the N-channel MOSFET (nMOS) are formed on a p-type substrate: nMOS enrichment (enhancement) or normally OFF and nMOS depletion (exhaustion) or normally on. The p-channel MOSFET (pMOS), instead, are constructed on an n-type substrate: pMOS enrichment (enhancement) or normally off and pMOS depletion (exhaustion) or normally on [2, 3].

The operation of the MOSFET Enhancement, or e-MOSFET, can be described using its characteristic curves described in Chap. 1 and reported in Fig. 3.1. When the input voltage (V_{in}) (gate) is zero, the MOSFET does not conduct and the output voltage is equal to the supply voltage. In this way, the MOSFET is "fully-OFF" and is in its cut-off region. When input voltage is HIGH or equal to supply voltage (V_{dd}), the operating point of the MOSFET moves to the point A along the load line.

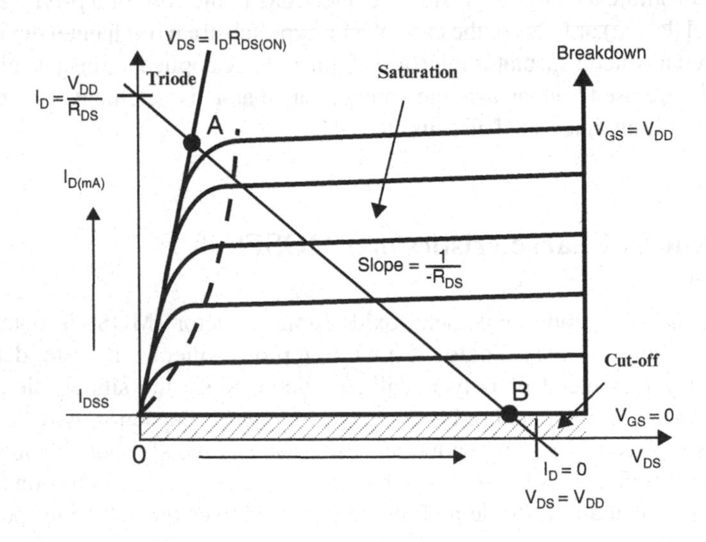

Fig. 3.1 Id–Vd characteristic curve

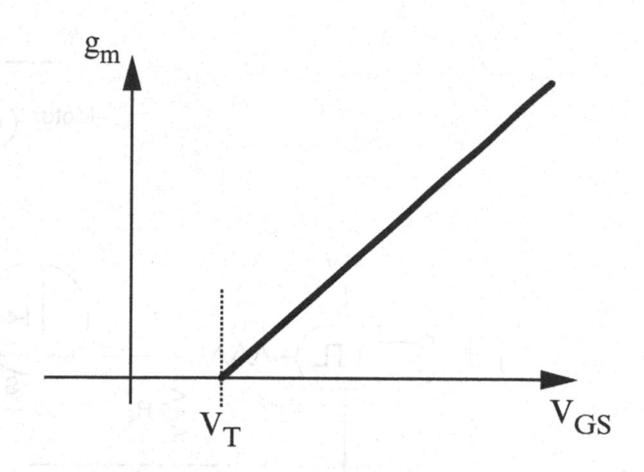

Fig. 3.2 Transconductance gm

The drain current I_D increases to its maximum value due to reduction of the channel resistance, and becomes a constant value independent from V_{dd} and depending only on V_{GS}. Therefore, the device behaves as a closed switch with a minimum channel resistance (R_{ds}). Similarly, when V_{IN} is low or reduced to zero, the operating point of the MOSFET moves from point A to point B along the load line. The channel resistance is very high so that the transistor behaves like an open circuit and no current flows through the channel. If the gate voltage of the MOSFET alternates between two values, high and low, the MOSFET behaves like a sensor "single-throw single-pole" (SPST) in the solid state. Referring to the $I_d - Vd$ characteristic of a MOSFET is also possible to define a set of parameters. While the transconductance gm (Fig. 3.2) gives the variations of the current I_{ds} (output) as a function of the voltage V_{gs}, there is another fundamental parameter that takes the name of output conductance. It expresses the variation of the current I_{ds} as a function of voltage V_{ds}. All two parameters are used to characterize the MOSFET in the amplifier configuration. Other parameters that characterize a MOSFET are the following: R_{ds} (on), the minimum resistance in conduction mode and can vary from a few ohms to a few milliohms; V_{gs} (th) or V_T, the voltage applied to the gate to conduct the Mosfet, is usually greater than 4 V; Q_{GD}, the minimum amount of energy necessary to the gate to switch on the MOSFET; t_d (on) is the time taken to charge the input capacitance of the device before drain current conduction can start; t_d (off) is the time taken to discharge the capacitance after switched off [4].

3.3 Mosfet Power Control

By reason of the very high gate resistance (input), its high switching speed and ease of manage, the MOSFET is ideal for layout with operational amplifiers or standard logic gates. In this case, the input voltage of the gate-source should be

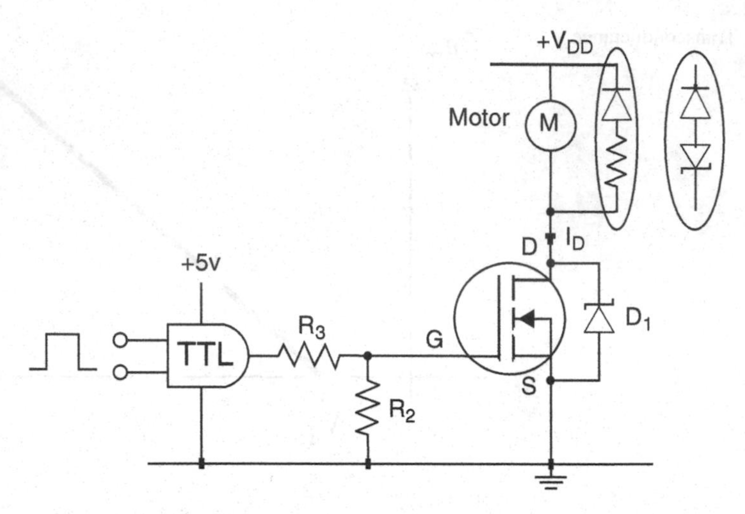

Fig. 3.3 Example of layout for power control

chosen properly, the device must have a low value of R_{ds} (on) in proportion to the input voltage. The power MOSFETs can be used to control the movement of the DC motors or stepper motors brushless directly from the logic computer or by using the PWM modulation. As a DC motor offers high starting torque proportional to the current, the MOSFET in PWM mode can be used as a speed regulator which provides a smoother and quieter operation of the engine. Since the engine load is inductive, a simple diode is connected across the load to dissipate any electromotive force generated by the motor when the MOSFET is in the "OFF" state. A network locking formed by a zener diode in series with the diode can also be used to permit faster switching and an improved control of the peak inverse voltage and the drop-out time (Fig. 3.3).

3.4 Stage of Amplification

The main stages of amplifiers using Mosfet can be described in three configurations: common source, common drain, and common gate. In a common-source amplifier, for example, the input signal is the signal applied to the gate, extracted in output from the drain as current or voltage with respect to ground. The stage common-source has input resistance very high and aids in the functioning as transconductor and fits perfectly also as a voltage amplifier. This simply means that by reason of high output resistance, the voltage gain (he voltage ratio of the output signal and the input level) depends on the load resistance. This dependence on the load resistance limits its usefulness as a voltage generic amplifier. In a common drain or source amplifier is applied the input signal with respect to ground on the gate terminal of the MOSFET. The main application is related to the buffer, because its input resistance is extremely high, while its output resistance is reasonably low. Unlike a voltage buffer ideal, a source follower MOSFET (common-drain amplifier) provides a gain that is always less than one. Although the common source is capable of substantial power gains of the signal, the voltage gain limits its utility in the application of small signal. The input port of a common gate amplifier has a relatively small input resistance and applies the input signal in current. The output response to the input current applied is significantly extracted as a current signal and the gain is always less than unity. The common gate amplifier can be represented as the dual tracker source or source follower [1, 5].

3.4.1 Common Source

A common-source amplifier is a topology typically used as a voltage amplifier or transconductance. As transconductance amplifier, the input voltage is seen as a modulation of the current to the load. As a voltage amplifier, instead, the input voltage modulates the amount of current that flows through the MOSFET. However, the output resistance of the device is not high enough for a reasonable transconductance amplifier (ideally infinite), or low enough for a discrete voltage amplifier (ideally zero). Another serious drawback is the limited high frequency response characteristic of the amplifier. Therefore, the output is often routed through a voltage follower (common-drain stage, C_D) or a current follower (common gate, C_G) to obtain the output characteristics and frequency response more favorable. The combination $C_S - C_G$ is called a cascode amplifier. The bandwidth of the common source amplifier tends to be low, because of the high capacitance due to the effect Miller. Figure 3.4 shows a common source amplifier with an active load. Figure 3.5 shows the relevant circuit for small signals when it is added a resistor load R_L to

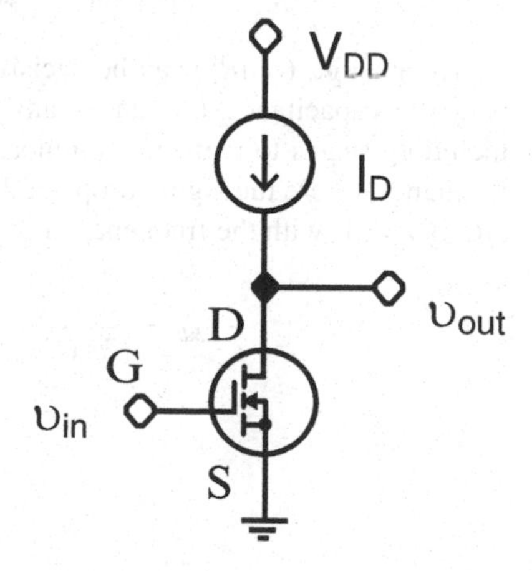

Fig. 3.4 Example of common source amplifier (base)

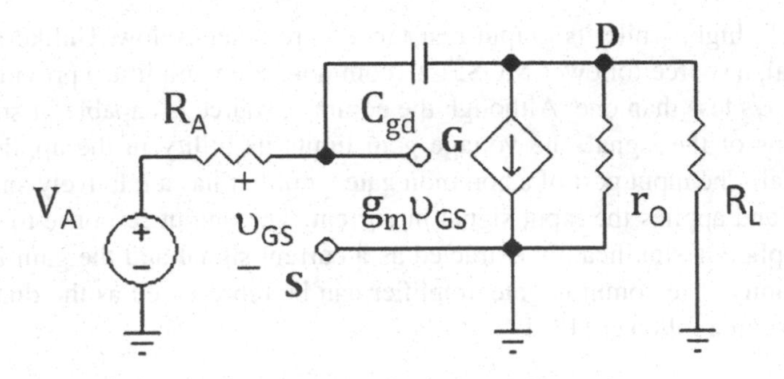

Fig. 3.5 Small signal circuit (common source)

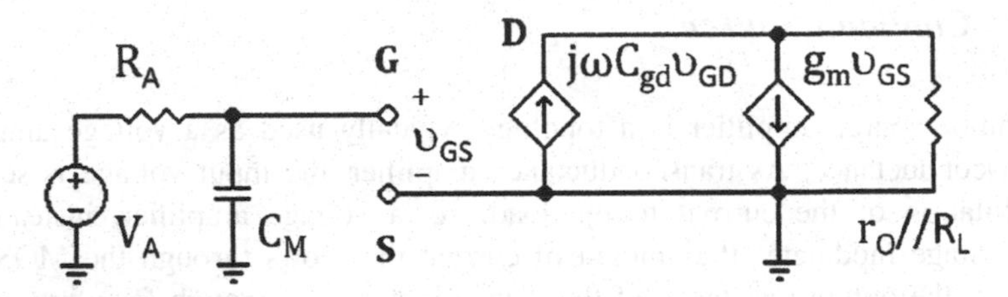

Fig. 3.6 Small signal circuit (common source) with Miller

the output node. The limitation on bandwidth in this circuit derives mainly from the coupling of the parasitic capacitance C_{gd} between gate and drain and the series resistance R_A.

By using the theorem of Miller, the circuit of Fig. 3.5 is transformed into Fig. 3.6, where is shown the capacity of Miller C_M on the input side of the circuit given by the following expression:

$$C_m = C_{gd}(1 + g_m(r_O||R_L)) \tag{3.1}$$

The gain g_m $(r_O||R_L)$ can be decisive in the frequency response even for a small parasitic capacitance C_{gd} and many "tricks" are used to counteract this effect, including stages to realize a common gate circuit (cascode). The bandwidth is the frequency where the signal drops of 3 dB (or 1/2). A reduction of 1/2 occurs when $\omega C_M R_A = 1$, with the frequency at 3 dB is expressed by the following formula:

$$f_{3db} = \frac{1}{2\pi(C_{gs} + C_{gd}(1 + g_m(r_O||R_L)))} \tag{3.2}$$

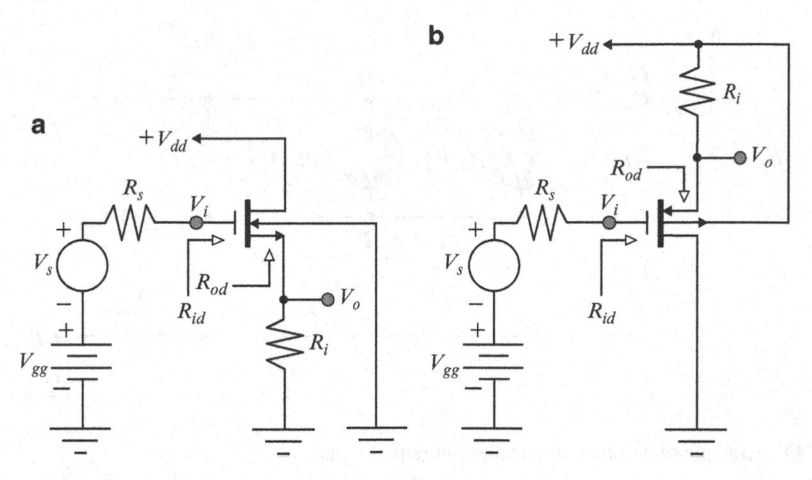

Fig. 3.7 Example of common drain amplifier: (**a**) NMOS, (**b**) PMOS

3.4.2 Common Drain

A common-drain amplifier is one of the three fundamental stages, typically used as voltage buffer. The fundamental structure of a common drain NMOS amplifier is drawn in Fig. 3.7. In this circuit (NMOS) the gate terminal of the MOS transistor serves as an input, the source is the output and the drain is common to both pins (input and output). The small circuit for common drain stage is visualized in Fig. 3.8. The output response, which is traditionally a signal voltage, is extracted from the source terminal. Unlike the common source, the drain common is rarely used as independent amplifier. It is often used in combination with a common source amplifier that has the goal of providing the gain to a highly capacitive load at high frequencies of the signal or, more generally, to any load of low impedance. As such, the source follower serves as a buffer inserted between the output port of a common source amplifier and the load [6–8].

3.4.3 Common Gate

The last of the three amplification stages of MOSFET technology is the common gate amplifier. This type of amplifier (Figs. 3.9 and 3.10) is typically used as a current buffer or voltage amplifier. In this circuit, the source terminal serves as an input, the drain is the output, and the gate is connected to ground. The layout is useful, for example, in CMOS RF receivers, especially when operating near the limits of the frequency of the FET/MOSFET; it is preferable, also, because of the ease of impedance matching and has a lower noise than other configurations [6].

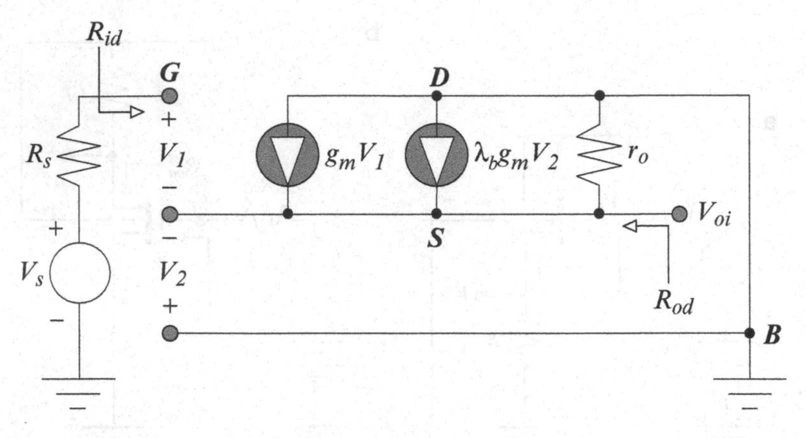

Fig. 3.8 Example of common drain amplifier (small signal)

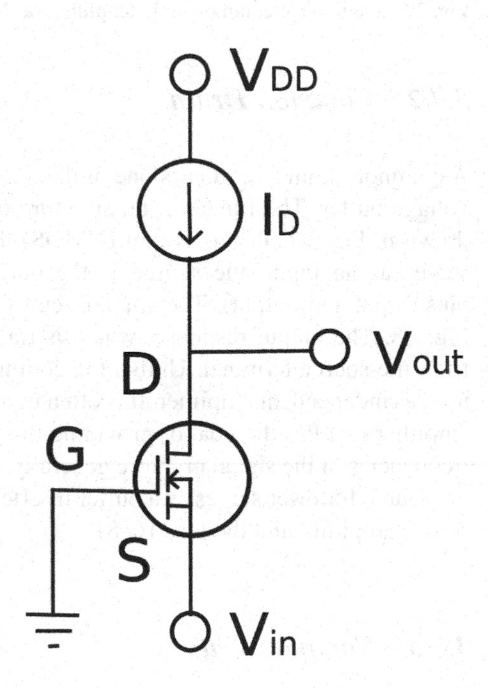

Fig. 3.9 Common gate amplifier

Fig. 3.10 Common gate amplifier (small signal)

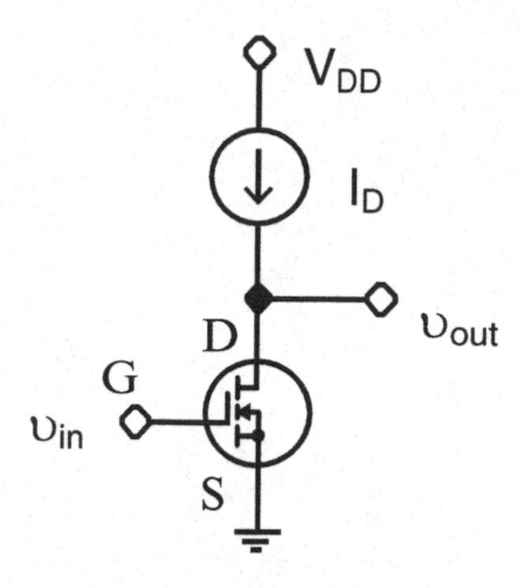

References

1. Razavi B (2002) Design of analog CMOS integrated circuits. McGraw-Hill, New York
2. Di Paolo Emilio M (2013) Data Acquisition system from fundamentals to applied design Springer, New York
3. Razavi B (2008) Fundamentals of microelectronics. Wiley, UK
4. Neudeck GW (1989) Modular series on solid state devices: volume III: the bipolar junction transistor, 2nd edn. Prentice Hall, Englewood Cliffs
5. Sedra AS, Smith KC (2013) Microelectronic circuits. Oxford University Press, New York
6. Razavi B (2002) Design of integrated circuits for optical communications. McGraw-Hill, New York
7. Hurst PJ (2001) Analysis and design of analog integrated circuits. Wiley, New York
8. Fonstad CG (1994) Microelectronic device and circuits. McGraw-Hill, New York

Chapter 4
Operational Amplifier

Abstract An operational amplifier (called op-amp) is a specially designed amplifier in bipolar or CMOS (or BiCMOS). It is used as generic "black box" building blocks in much analog electronic design: Amplification, Analog filtering, Buffering, Threshold detection. The op-amp can also be represented as a dependent voltage source, having an output impedance and input impedance. The input impedance is so high that no current can flow between the input terminals, but the output impedance is very low.

4.1 Introduction

An operational amplifier (called op-amp) is a specially designed amplifier in microelectronics with the following typical characteristics (Figs. 4.1 and 4.2):

- Very high gain (10,000 to 1,000,000)
- Differential input
- Very high (assumed infinite) input impedance
- Single ended output
- Very low output impedance
- Linear behavior

The circuitry that makes up an op-amp consists of transistors, resistors, diodes, and a couple capacitors. In general, these components are combined to achieve within the op-amp two stages of differential amplifiers and a common-collector amplifier. In an effort to simplify the operational amplifier, one must not forget that the internal circuitry of an op-amp is more than just a black box. The amplifier's differential inputs consist of a non-inverting input (+) with voltage V+ and an inverting input (−) with voltage V−; ideally, the op-amp amplifies only the difference in voltage between the two, which is called the differential input voltage [1, 2].

M. Di Paolo Emilio, *Microelectronics*, DOI 10.1007/978-3-319-22545-6_4

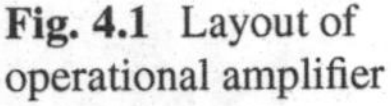

Fig. 4.1 Layout of operational amplifier

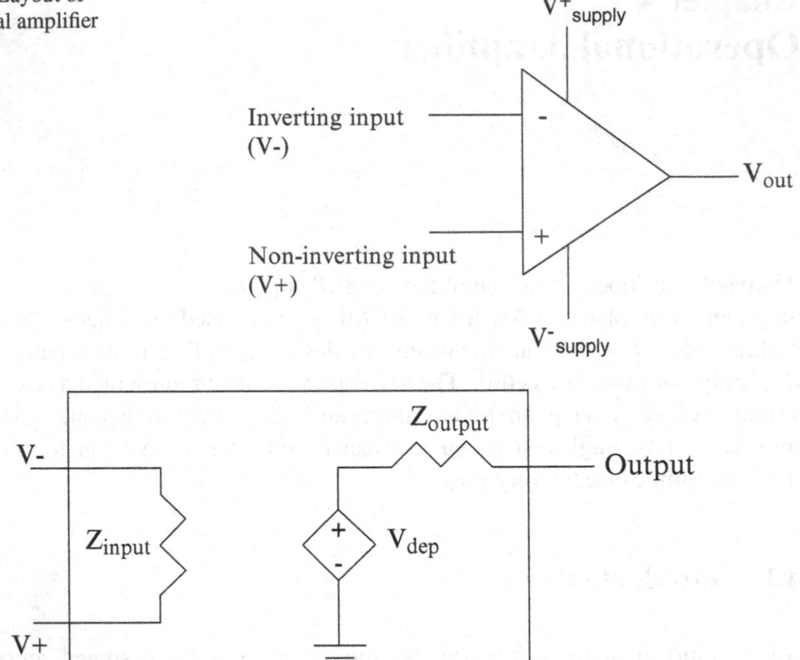

Fig. 4.2 Equivalent layout of op-amp

4.2 Main Configuration

An op-amp can use negative feedback to set the closed-loop gain as a function of the circuit external elements (resistors), independent of the op-amp gain, as long as the internal op-amp gain is very high. In Fig. 4.3 is shown an ideal op-amp in a non-inverting configuration with negative feedback provided by voltage divider R1, R2.

The op-amp with negative feedback forces the two inputs v+ and v− to have the same voltage, even though no current flows into either input. This is sometimes called a "virtual short." As long as the op-amp stays in its linear region, the output will change up or down until v− is almost equal to v+. If v_{IN} is raised, v_{OUT} will increase just enough so that v− (tapped from the voltage divider) increases to be equal to v+ (= v_{IN}). The negative feedback forces the virtual short condition to occur.

An op-amp in the inverting configuration (with negative feedback) is shown in Fig. 4.4. Feedback is from v_{OUT} to v− through resistor R2. v_{IN} comes in to the v− terminal via resistor R1 and v+ is connected to ground [3, 4].

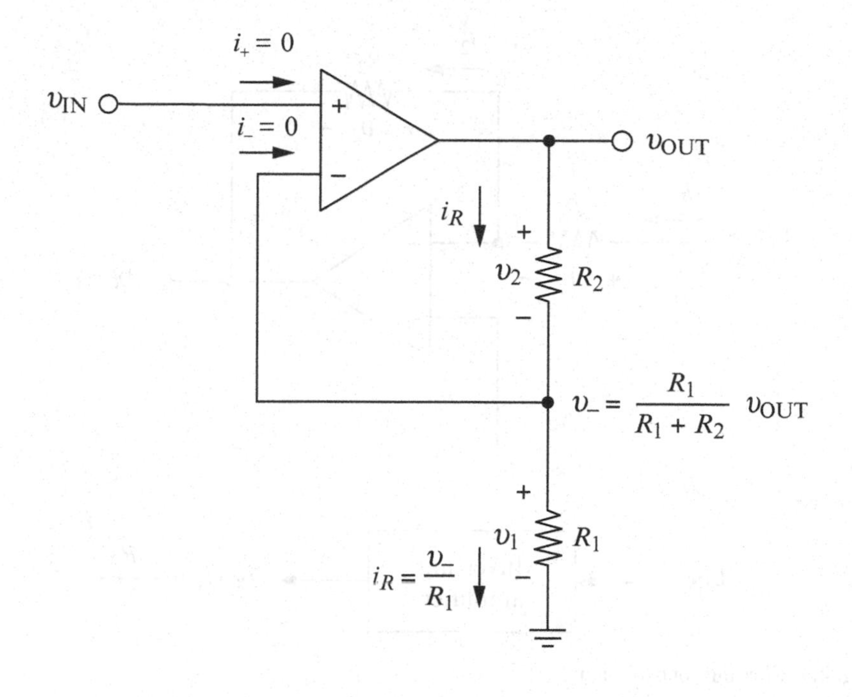

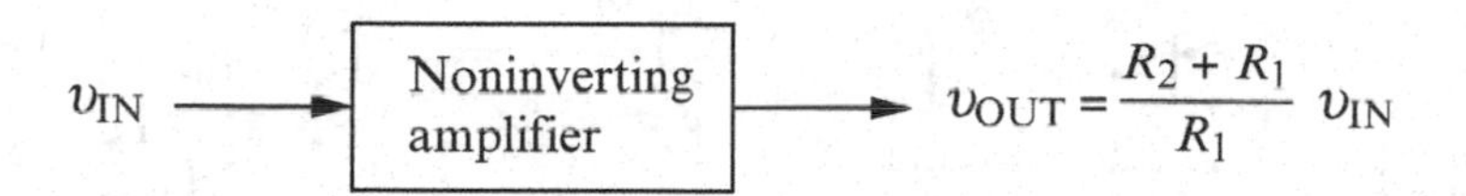

Fig. 4.3 Non-inverting configuration

4.3 Real Operational Amplifier

Real operational amplifiers suffer from several non-ideal effects (Figs. 4.5 and 4.6). Open-loop gain is infinite in the ideal operational amplifier, but typical devices exhibit open-loop DC gain ranging from 100,000 to over 1 million. The differential input impedance of the operational amplifier is defined as the impedance between its two inputs; the common-mode input impedance is the impedance from each input to ground.Due to biasing requirements or leakage, a small amount of current (typically about 10 nA for bipolar op-amps, 10 pA for JFET input stages, and only a few pA for MOSFET input stages) flows into the inputs. A perfect operational amplifier amplifies only the voltage difference between its two inputs, completely rejecting all voltages that are common to both. However, it is never perfect, and the standard measure of this defect is called the common-mode rejection ratio (denoted CMRR). Amplifiers generate random voltage at the output even when there is no signal applied. This can be due to thermal noise and flicker noise of

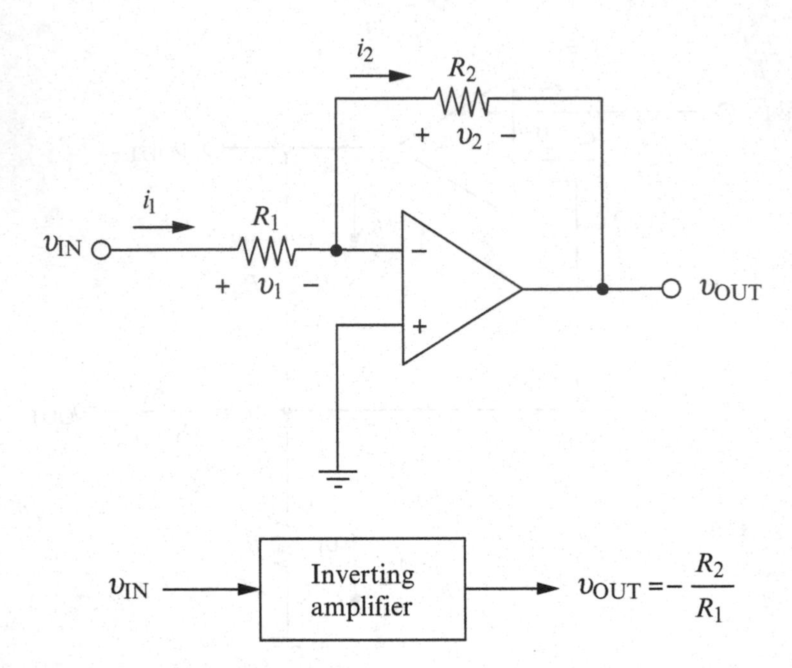

Fig. 4.4 Inverting configuration

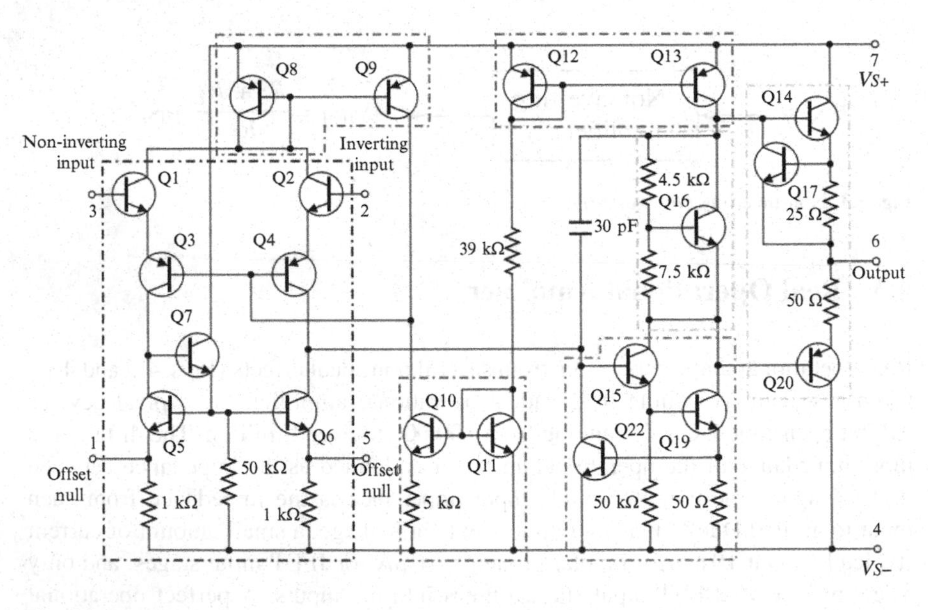

Fig. 4.5 Internal configuration of commercial op-amp (ua741)

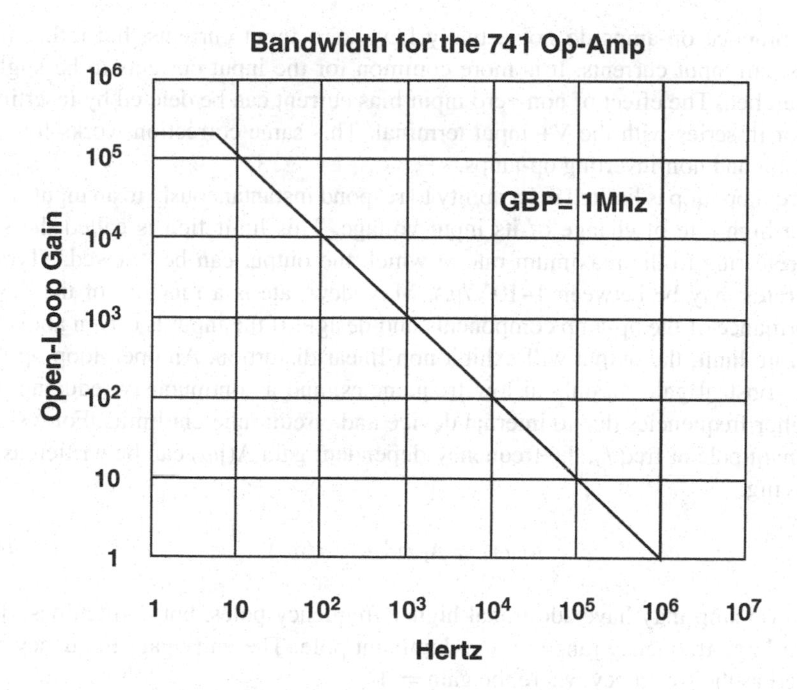

Fig. 4.6 Example of frequency response of op-amp

the devices. For applications with high gain or high bandwidth, noise becomes a very important consideration. The amplifier's output voltage reaches its maximum rate of change, the slew rate, usually specified in volts per microsecond. Modern high speed op-amps can have slew rates in excess of 5 kV/μs. However, it is more common for op-amps to have slew rates in the range 5–100 V/μs.

4.4 Electrical Characteristics

Real op-amps differ from the ideal model in various aspects. In actual practice an op-amp circuit will saturate at somewhat lower than V_{POS} and higher than V_{NEG}, due to internal voltage drops in the design (Output Saturation Voltage). In the perfect amplifier, there would be no input offset voltage. However, it exists in actual op-amps because of imperfections in the differential amplifier that constitutes the input stage of the vast majority of these devices. Input offset voltage creates two problems: First, the amplifier output will go into saturation if it is operated without negative feedback, second, in a closed loop, the input offset voltage is amplified along with the signal and this may pose a problem in the amplification of small input signal [5].

In practice op-amps do not actually have zero input currents, but rather have very small input currents. It is more common for the input currents to be slightly mismatched. The effect of non-zero input bias current can be deleted by inserting a resistor in series with the V+ input terminal. This same correction works for both inverting and non-inverting op-amps.

A real op-amp is limited in its ability to respond instantaneously to an input signal with a high rate of change of its input voltage. This limitation is called the slew rate, referring to the maximum rate at which the output can be "slewed." Typical slew rates may be between 1–10 V/μs. Max slew rate is a function of the device performance of the op-amp components and design. If the input is driven above the slew rate limit, the output will exhibit non-linear distortion. An open-loop op-amp has a constant gain A_0 only at low frequencies, and a continuously reducing gain at higher frequencies due to internal device and circuit inherent limits. For a single dominant pole at freq f_p, the frequency-dependent gain A(jω) can be written as the following:

$$A(j\omega) = A_0/(1 + j\omega/\omega_p) \quad (4.1)$$

An op-amp may have additional higher frequency poles, but is often described over a large frequency range by the dominant pole. The unity gain frequency f_0 is defined as the frequency where the gain = 1.

The gain of a typical op-amp is inversely proportional to frequency and is characterized by its gain bandwidth product (GBWP). For example, an op-amp with a GBWP of 1 MHz would have a gain of 5 at 200 kHz, and a gain of 1 at 1 MHz. A typical op-amp contains circuitry to limit the output current to a specified maximum in order to protect the output stage from damage. If a low value load impedance is utilized, the output current limit may be reached before the output saturates at the rail voltage, forcing the op-amp to lower gain [6].

4.5 Circuits with Operational Amplifier

The op-amp configuration shown in Fig. 4.7 is a voltage-follower often used as a buffer amplifier. Output is connected directly to negative input (negative feedback). Since v+ = v− = v_{IN}, and v_{OUT} = v−, it's possible to observe by inspection that the closed-loop gain $A_0 = 1$.

The differential amplifier shown in Fig. 4.8 combines both the inverting and non-inverting op-amps into one circuit. The Differential Amplifier circuit is a very useful op-amp circuit and by adding more resistors in parallel with the input resistors R_1 and R_f, the resultant circuit can be made to either "Add" or "Subtract" the voltages applied to their respective inputs. One of the most common ways of doing this is to connect a Resistive Bridge commonly called a Wheatstone Bridge to the input of the amplifier as shown below. Some applications, such as an oscilloscope input, require differential amplification with extremely high input resistance such as circuit shown

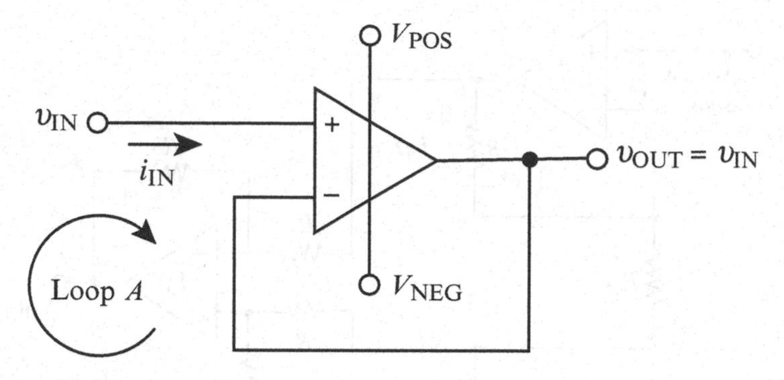

Fig. 4.7 Voltage-follower

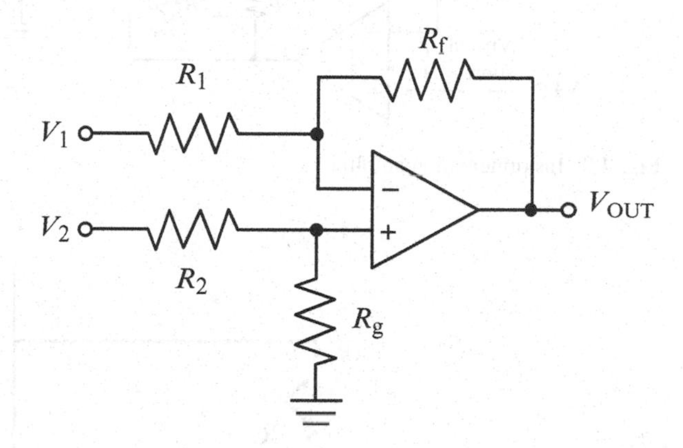

Fig. 4.8 Differential amplifier

in Fig. 4.9. A3 is a standard difference op-amp with differential gain R_2/R_1. A1 and A2 are additional op-amps with extremely high input resistances at v_1 and v_2 (input currents = 0).

Instrumentation amplifiers are a combination of three op-amps that are typically grouped into two stages. The first two op-amps comprise the first stage and each is a non-inverting amplificr. Thc sccond stagc is a differential amplifier that may or may not have unity gain. An instrumentation amplifier is beneficial for several reasons:

- High input impedance, unlike the lower input impedance of a differential amplifier by itself.
- High CMRR
- Good for smaller, insignificant input signals
- Gain of the non-inverting amplifiers can be varied by the rheostat(RC).

A comparator is a device that can be used to compare an input voltage to a reference voltage. The circuitry for comparators can vary in design and thereby vary in results. This configuration uses a voltage divider to create the reference voltage (Fig. 4.10).

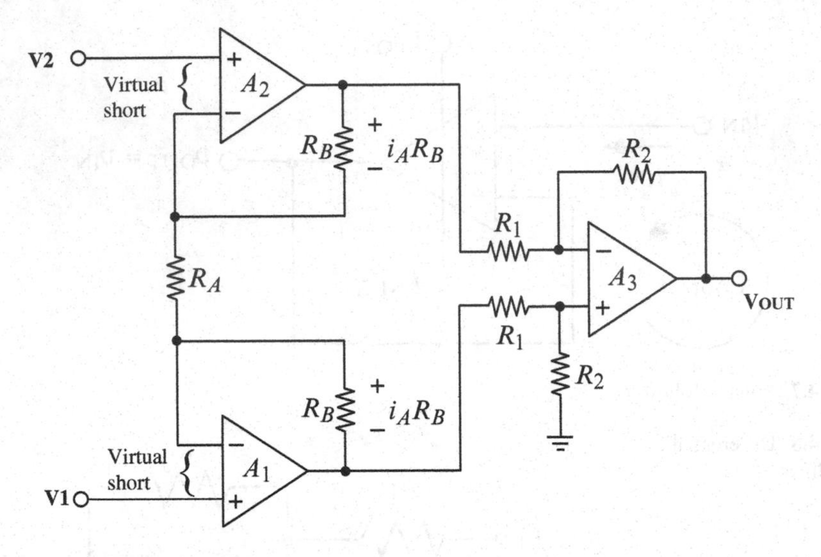

Fig. 4.9 Instrumentation amplifier

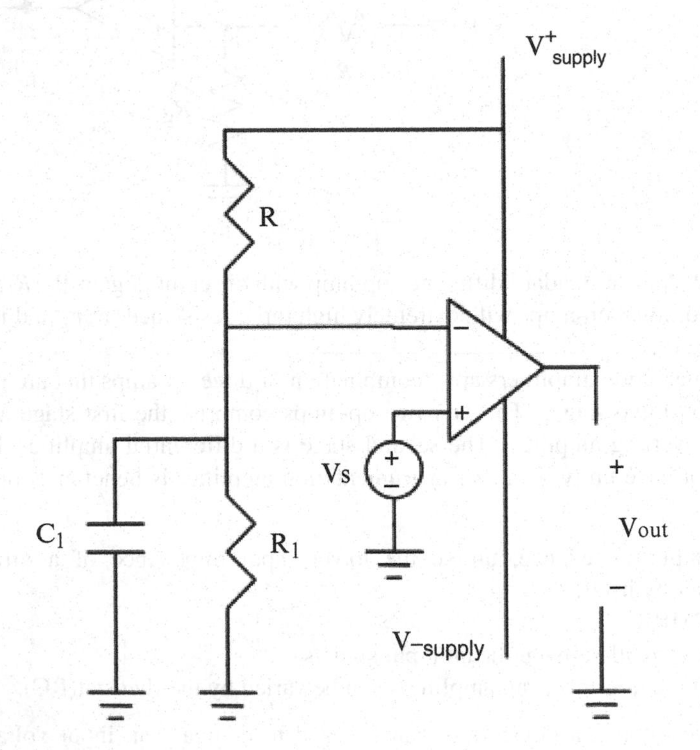

Fig. 4.10 Comparator

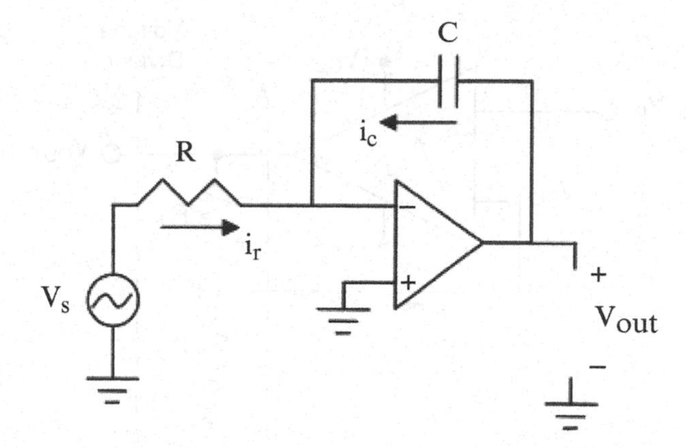

Fig. 4.11 Integrator circuit

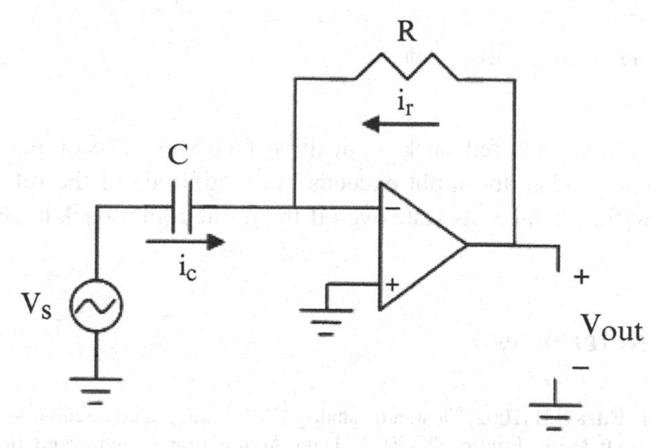

Fig. 4.12 Differentiator circuit

The subject of calculus involves exercises in differentiation and integration of which most students struggle in varying degrees to understand. Well, the amazing versatility and value of the op-amp can be seen in its ability to perform integration and differentiation. Integrators and differentiators, as they are called, are very similar and their circuits are simple to draw. The use of a capacitor is what makes the complex mathematical process possible. In Fig. 4.11 is shown the integrator where the output will be equal to the integral of the input, as long as the op-amp remains in its linear region. Differentiation is the counterpart to integration and by simply switching the location of the resistor (R) and capacitor (C), a differentiator circuit can be formed. Since a capacitor does not allow dc current to pass through it, the voltage sources associated with the integrator and differentiator circuits are ac sources (Fig. 4.12).

The open-loop comparator is very susceptible to noise on the input. Noise may cause it to jump erratically from + rail to − rail voltages. The Schmitt Trigger circuit (Fig. 4.13) solves this problem by using positive feedback. It is a comparator circuit in which the reference voltage is derived from a divided fraction of the output

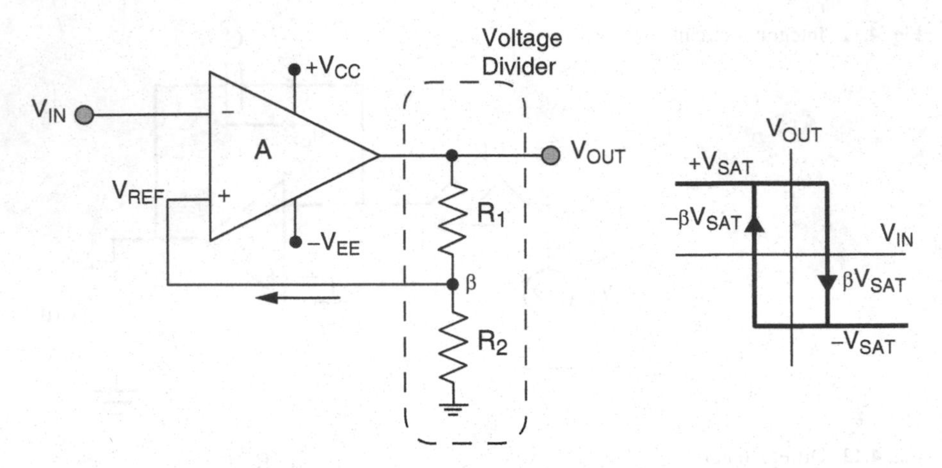

Fig. 4.13 Trigger Schmitt

voltage, and fed back as positive feedback. The output is forced to either V_{sat} or $-V_{sat}$ when the input exceeds the magnitude of the reference voltage. The circuit will remember its state even if the input comes back to zero (memory) [7, 8].

References

1. Razavi B (2002) Design of analog CMOS integrated circuits. McGraw-Hill, New York
2. Di Paolo Emilio M (2013) Data Acquisition system from fundamentals to applied design. Springer, New York
3. Razavi B (2008) Fundamentals of microelectronics. Wiley, Hoboken
4. Wikipedia (2008) http://en.wikipedia.org/wiki/Operational_amplifier
5. Sedra AS, Smith KC (2013) Microelectronic circuits. Oxford University Press, New York
6. Razavi B (2002) Design of integrated circuits for optical communications. McGraw-Hill, New York
7. Gray PR, Hurst P, Meyer RG, Lewis S (2001) Analysis and design of analog integrated circuits. Wiley, New York
8. Fonstad CG (1994) Microelectronic device and circuits. McGraw-Hill, New York

Chapter 5
Design PCB

Abstract The design of a printed circuit board (PCB) is a very important task to realize electronic prototypes efficiently from both an operational point of view and commercial. Basically, in the microelectronics applications, the design of the PCB plays a key role. The circuits are embedded in various types and sizes, in relation to the type of microprocessor, component, and operating system, above all, the complexity of the software can be various from a few hundred of bytes to several megabytes of code.

5.1 Materials for Printed Circuits

A printed circuit board (PCB) is a set of copper tracks suitably drawn on an insulating support and used to connect the components that constitute the electronic circuit. The base material is formed by the copper sheet of appropriate dimensions and the insulating part that can be in different types according to the performance; the classic Bakelite is the insulating material which is cheaper and has less performance.

The PCB (Fig. 5.1) are made by assembling thin dielectric layers which are made with electrically conductive material:

- One or more rigid or flexible substrates act as insulation.
- Layers on copper levels conduct the electric connections.
- Holes conductors ("vias") cross-connect different layers.
- A painted surface is used to protect the slopes from oxidation and facilitate welding ("solder").
- Screen printing ("Silkscreen") is to mark the location of the components on the board.

For the choice of the dielectric material of the substrate, there are some factors needed to be considered:

Coefficient of Thermal Expansion (CTE): The CTE is the tendency of the material to change in volume in response to a variation of temperature. When a substance is heated, the particles begin to move more and thus usually maintain a higher average separation.

M. Di Paolo Emilio, *Microelectronics*, DOI 10.1007/978-3-319-22545-6_5

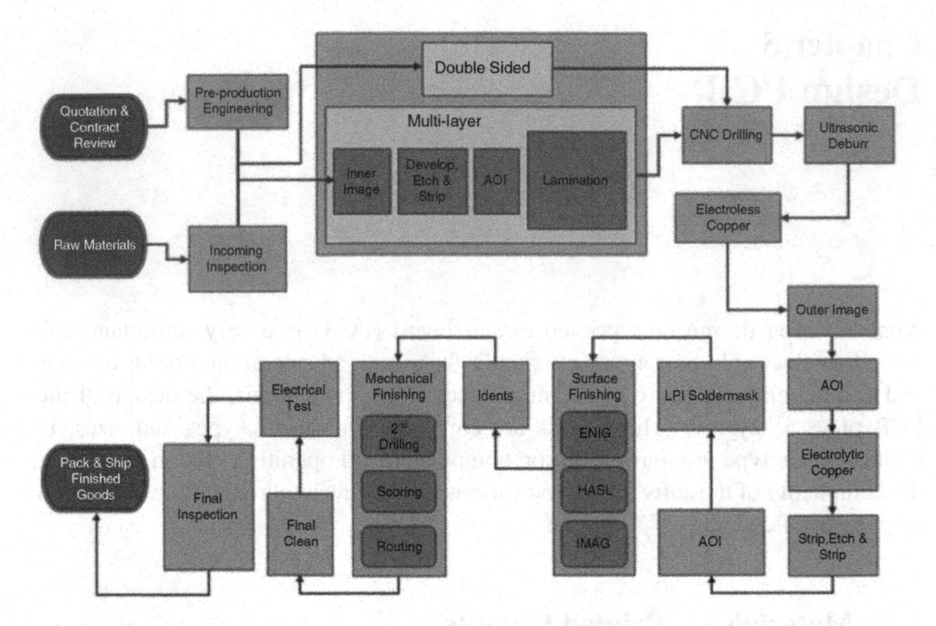

Fig. 5.1 PCB design flow

Glass Transition Temperature (T_g): The glass transition is a property of only the amorphous portion of a semi-crystalline solid. The crystalline portion remains crystalline during the glass transition.

Thermal conductivity: The ratio between the heat flow and the temperature gradient. It represents the measure of the ability of a substance to conduct the heat, generally depends on the nature of the material itself but not on its shape.

Rigidity mechanics: The ability of a body to oppose an elastic deformation due to an applied force.

FR4 substrate is most commonly used to make circuit boards; formed from glass fibers and copper joined by an epoxy resin. It is the most common material used in electronics and mechanics, lightweight with high resistance to mechanical stress, and good resistance to thermal shock of short duration.

FR-4 is acceptable for signals up to about 2 GHz, depending on the application in used, in addition, for the production of insulating and structural components. If the same board of FR-4 delivers more high-frequency signals, it would cause power loss and interference increase. Other materials provide superior electrical characteristics at higher frequencies.

Substrates commonly used are the following:

- Polymide/fiberglass;
- FR4 (Fig. 5.2);
- KAPTON (Fig. 5.3): flexible and lightweight, used for specific applications (displays, keyboards).

Fig. 5.2 FR4

Fig. 5.3 Kapton

The polyamides are known for its thermal stability, chemical resistance, and excellent mechanical properties. The compounds are reinforced with glass fiber or graphite to a high flexural strength up to 345 MPa. Kapton (Fig. 5.3) is a polyamide film developed by DuPont which is enabled to remain stable for temperatures ranging from −269 °C to +400 °C. It's used as well as in flexible printed circuits even in the astronauts space and suits to ensure thermal protection.

For the realization of a PCB performing in different steps of manufacturing, particularly, the first phase of the setup consists in the choice of materials, processes, and requirements. The phase can be defined in the following points:

- It is called the quality and reliability of the card, with reference to the specifications.
- The materials are chosen for the core and the layers of materials referring to the IPC- 4101.
- Stack-Up is defined layers (core stack-up or foil stack-up).

Other materials which may be used in PCB are:

- CEM-1. It consists of two layers of woven glass fiber with a layer of paper in the middle; both of these materials are impregnated with an epoxy resin. It has the best electrical and physical characteristics compared to the FR-1 and FR-2.
- CEM-3. It consists of two layers of woven glass fiber, with a non-woven glass fiber in the middle, both impregnated with an epoxy resin.

In addition to the choice of the material, the PCB coating is of fundamental importance; coating is made through the special protective resins and consists in the assembling in the PCB with a resin film that crystallizes at a certain temperature. It forms a single body with the substrate and welded components, isolates them from the external elements.

The main goal is to obtain a barrier of insulation for electronic components, so that the PCB can run in any operating mode; in this way, the damage is limited due to mechanical stress, humidity, and more.

The resins used are in acrylic-based and silicone, which is selected according to the operating environment.

The main characteristics of the most common resins used can be listed in the following:

- Epoxy resins:
  - Good moisture and temperature resistance;
  - Thermal shock resistance.
- Resins puliuretaniche:
  - Moisture and thermal shock resistance;
  - Low temperature resistance;
- Silicone resins:
  - Excellent electrical conductivity;
  - Non-economic;

Polyurethane resins are preferred when PCBs are formed by relatively delicate components such as ferrites that might be interfering RF signals. Silicone resins, however, are very expensive for its excellent electrical capacity and are used when the PCB operates at high temperatures (higher than 180 °C) [1–4].

5.2 Electrical Insulation on PCB

The electrical insulation is a condition in which between two points, with a different potential, there is a movement of continuous current. At the physical level, there is a shift of electrical charge from one point to another while the electrical power is exchanged by electromagnetic inductive or capacitive phenomena. The magnitude of the electric type that urges an insulating material, defined as stress dielectric, is the electric field E [V/m]. When the stress is too high, the dielectric could cause a temporary or permanent damage (depends on the material type) of the insulation that impairs the functionality of the machine to which the insulator shall be included.

It is defined as E_r dielectric strength [V/m] of a material the maximum value of the dielectric stress which may be applied without damage.

The dielectric strength (Fig. 5.4) is essentially a random parameter which must be done on a statistical basis. It also depends on several factors:

- Waveform and duration of the applied voltage;
- Geometry of the electrodes and insulator;
- Physic-chemical characteristics of the material;
- The presence of impurities in the material (moisture, gas, waste, etc.);
- Thermal and mechanical stresses applied to the material.

The electric field in an insulating material is determined by the applied voltage V and its geometry: $E = -\nabla V$ is proportional to the applied voltage, but in the homogeneous material, it is not determined by the permittivity of the material itself. The dimension of the dielectric insulating material consists in determining of the geometry so that the pressure does not cause the damage to the dielectric; this means that the size and shape of the material must be conformed with the dielectric stress that is lower than the dielectric strength: $E < E_r$.

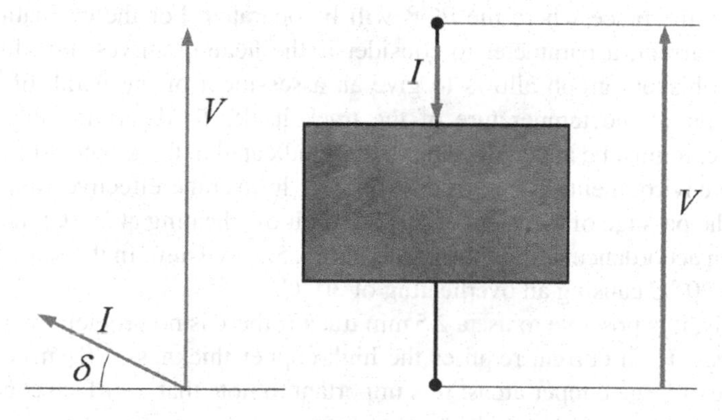

Fig. 5.4 Dielectric strength of insulation

Since the dielectric stress E is determined on the applied voltage V, it is obvious that the dimension must be done in relation to a well-determined value of the applied voltage.

It defines the level of insulation value of the applied voltage that determines the dimension of a dielectric insulating material.

The level of insulation is usually higher than the value of the nominal voltage, and this has the following characteristics:

- The size of the insulation must take consideration of the surges, long-lasting or transient that can stress the system under certain operating conditions or anomalies effect.
- In connection with the reliability is intended to provide to the equipment: more security and highest level of isolation.

The correct sizing of isolation is verified with the tests in the laboratory by applying insulation test voltages which is provided by the insulation level; it usually indicates the test voltages that the machine or equipment must be ensured to remain undamaged.

A PCB must be isolated to ensure the protection level and, therefore FR-4 is preferred as one of the PCB materials, economic alternative solution could be as synthetic resin bonded paper (SRBP, FR-1, FR-2). Due to some of its physical-mechanical characteristics and the ability to keep data at high frequencies, flame and heat resistance, water absorption is less than the other materials, FR-4 is widely used for the high-end construction in the industry, military, and consumer electronic equipment. It is also compatible with the "ultra-high isolation" [ultra high vacuum (UHV)].

The realization of the tracks in the PCB must be complied with the certain protocols, particularly for the printed circuits at high voltage and/or current is required more tracks width and isolation distance. The choice of sizing must not only follow certain rules but also specify the environmental techniques for determining the place where the PCB will be operated. For the evaluation of the maximum current, a parameter to consider is the heating curves. For thicknesses of 35 μm whiskers graph allows to give an assessment of the width of the track as a function of the temperature of the track itself. To determine the surface's temperature, it must be added the temperature indicated in the graph of Fig. 5.5 with that of the environment. The current display is in average effective value (RMS). Consider the passage of a current of 10 A to limit overheating at 10 °C: the width of the track, in accordance with to the graph of Fig. 5.5, is 9 mm. In the same track can pass about 20 °C causing an overheating of 30 °C.

Similarly, it is possible to use a 2.5 mm track if there is no problem with the high temperatures. High current requires the high copper thickness (70 μm) of circuits and the use of large copper areas; it is important to note that a surface connected to a copper track allows to lower the temperature.

Sometimes the problem can be traced back to the resistance of the track (and therefore the voltage drop). To calculate this resistance may be used the classic law:

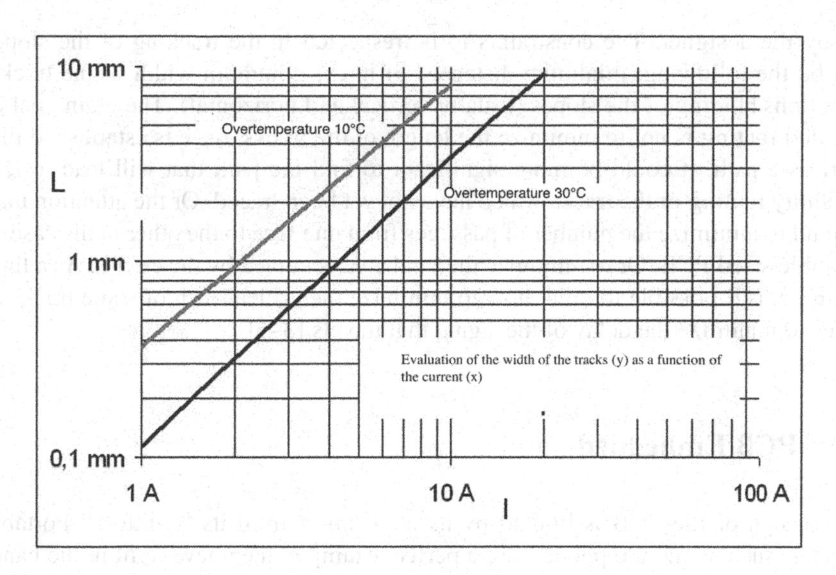

Fig. 5.5 Slopes size evaluation

$$R = \frac{\rho L}{A} \tag{5.1}$$

the resistivity of copper at 25 °C is about 0.018 μΩ m. A track with a 10 cm ($L = 0.1$ m) by the width of 1 mm and the thickness of 35 μm has a resistance of about 0.05 Ω. The parameters that allow an evaluation of the insulation distances are known as clearance and creep age. The first measured in air, the second measured in the following periphery of the insulating surface. A wide variety of environment (humidity, pressure, etc.) is very difficult to make an accurate assessment [3–6].

5.3 Routing PCB

The problem of routing for PCB is divided into two main phases: the global routing and detailed routing. The global routing includes the study on the optimal placement of components on the base and the identification of the paths that approximate the tracks will have to make from one component to another. This stage is very important because it allows splitting the base into various areas on which will go then to act the detailed routing.

The task is delicate because it is necessary to evaluate accurately the density of tracks and obstacles in each area so that it is then possible to complete the routing. The detailed routing is concerned to trace the connections between the components in each area identified from the global routing technology and respecting the constraints imposed on the project and optimizing some characteristics of the path set by the designer. The constraints to be respected in the tracking of the slopes can be the following: minimum distances of track, minimum width of the tracks, directions binding of the slopes (usually vertical and horizontal). The main goal of detailed routing is not to minimize the length of the tracks, once is established that there is a path, it could be more significant to find the path that will lead to less difficulty routing of the tracks which have not yet been traced. Or the attention may be paid to minimize the number of passages from one face to the other in the case of a double sided PCB. Or to minimize the total area occupied by the circuit. As a final example, is it possible to think how to minimize the critical path of some tracks in order to minimize the delay of the signal that travels [3–6].

5.4 PCB Embedded

The design of the PCB is limited by its application from its "comfort." Portable devices, such as mobile phones, are a perfect example, they have to fit in the hand, approach the ear to hear and equipped with a suitable size for easy use. These represent the minimum standards that have been changed since the introduction of the first GSM mobile phone. Still, consumers see higher capacity in the new handsets with the same size requirements.

There is an increase of active and passive electronic components, however, the space between these components is even more limited for electronic circuits design; then, PCBs become denser with additional components, tracks, and vias.

The number of discrete passive components is about 70–80 % of the total number and continues to grow much more. While the active components can be made in chips, the ideal space for placement of the discrete passive components becomes increasingly difficult to obtain. There are some considerations of which need to take into account:

- Reduce the length of the critical paths;
- Analog components separated physically from digital;
- Power components physically separated from other;
- Orientation of the components agrees with that of the slopes;
- Distribution and sizing of capacitors and filters for noise reduction (low and high frequency) or external coupling.

A component can be an embedded PCB with integrated active and passive components in a resin substrate. The design of a PCB with embedded monolithic ceramic capacitors is shown in Fig. 5.6. This scheme allows, as mentioned previously, an expansion of the structure components PCB and reducing the physical size. To facilitate the operation of these PCBs, a high quality power supply must be provided in order to absorb fluctuations in load and eliminate the noise during operation. For this reason, it is essential to reduce the inductance component by placing the passive components, such as a capacitor which is close to an IC in order to reduce the length of connection and to avoid unacceptable levels of noise that cause timing errors.

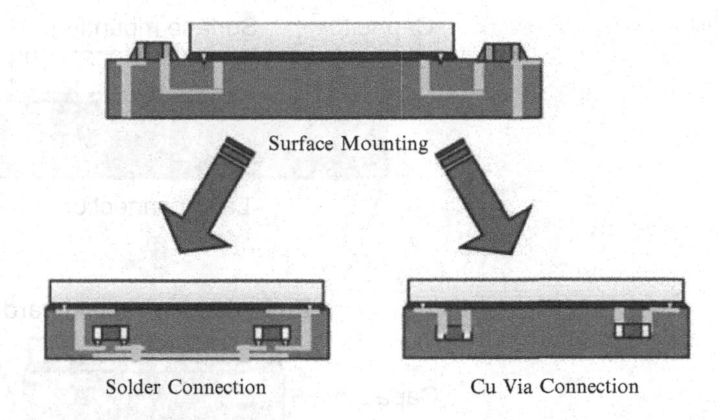

Fig. 5.6 Schematic of an embedded PCB

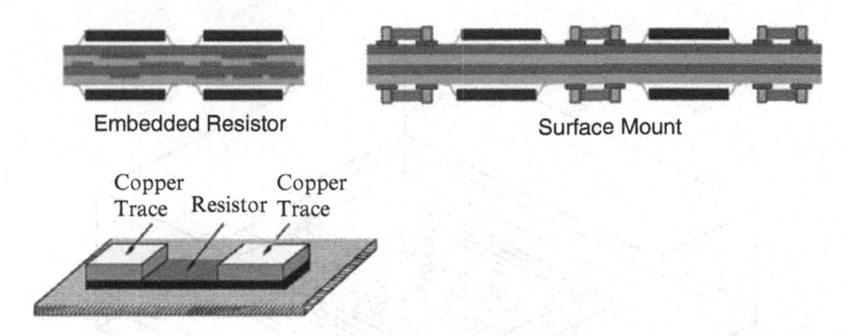

Fig. 5.7 Embedded resistor

The use of embedded passive components allows the use of the circuit at high frequencies. The linearity of the signals through these components also helps reduce the inductance to the surface and improve the electrical efficiency of the PCB. The reliability of the PCB would be improved through the reduction of the total number of welding joints. A welding joint is a point or edge where two or more pieces of metal or plastic are joined together. They are formed by welding two or more work pieces (metal or plastic) according to a particular geometry.

A resistor embedded resides on a single level, which can be any physical layer. It has two copper pads with resistive material between the electrodes. The shape of the resistance material may be a simple rectangle, or a shape shown in Fig. 5.7. In all cases, the resistance material overlaps the copper pads [3–6].

In general, the capacitors represent the highest percentage of components on a PCB. Today there are two techniques for the manufacture of embedded capacitors (Fig. 5.8): through the use of dielectric films and a process additive. The latter is a screening process which is created directly on a layer and is used in MCM and hybrids.

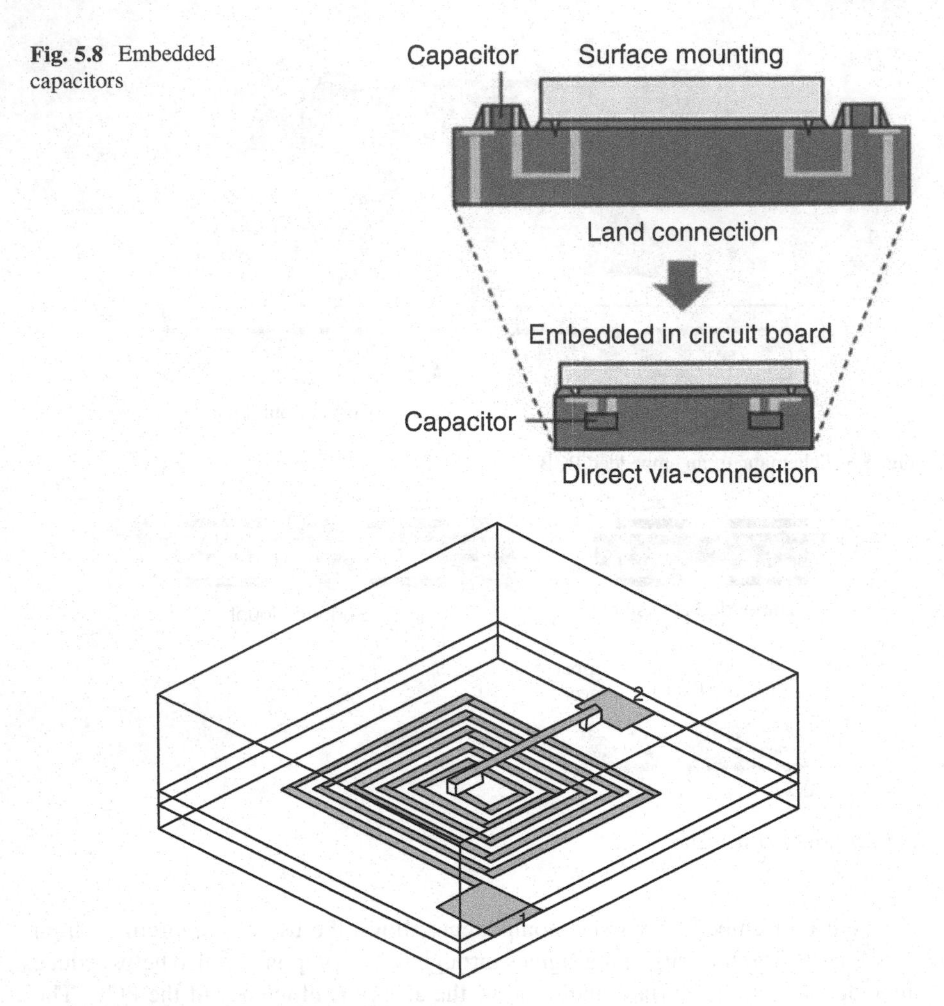

Fig. 5.8 Embedded capacitors

Fig. 5.9 Embedded inductors

Moreover, Embedded Inductors are realized by thin lines of spiral shape. The combinations of line width, number of turns, shape and length define the inductance. Figure 5.9 shows spiral lines of copper connected through a pad.

5.4.1 Design Guidelines

One of the most important points to understand the use of integrated passive components embedded is a manufacturing process but not an assembly process. Considering the design phase, this provides a great flexibility allowing the designers of PCBs to deal with the growing challenges of component density. Of course, such flexibility can be realized only if the Tools fully support embedded components. From the point of view of the design flow, the synthesis of the component must take place in the phases of positioning and interconnection of PCB design. This provides the optimal control of the design on the size of the component and the XY positioning on the layer during the definition of the critical aspects of the design. A summary of the appropriate component provides designers different physical options. Thus, two instances of resistance from 10 kW, for example, can be represented in a completely different way in a design for the best suit the design density. Interactive tools allow the designers to have an automatic synthesis forms and individual values or in groups selected in order to optimize the design density. Embedded discrete active components have different needs. They can be placed in a cavity exposed or embedded within the layers of the substrate. Compared to passive ones, have a much greater thickness, both for the physical component and assembly.

Moreover, the final stage of PCB design is the post-processing—the generation of outputs. This activity is typically done by the PCB designer. Design tool should provide Boolean AND, OR and XOR, which includes materials to determine what is included on the outputs of the plotter. These are used to create masks of appropriate materials and manufacturing processes. Additional outputs must support laser trimmer, XY positions, layer pads, and wire bonds. All these output capabilities are needed for each substrate.

Embedded components provide the opportunity for greater functionality, performance and density of PCBs at lower costs. However, these benefits can only be realized with the use of design tools, which includes and embraces the unique needs of the components. Tools for designing PCB must allow the flexibility to adopt the component at any point in the drawing so that it can be accomplished the best PCBs in terms of density [3–6].

References

1. Razavi B (2008) Fundamentals of microelectronics. Wiley, New York
2. Kang SM (1998) CMOS digital integrated circuits. McGraw-Hill, New York
3. Di Paolo Emilio M (2013) Data acquisition system from fundamentals to applied design. Springer, New York
4. Valdes-Perez FE (2009) Microcontrollers, fundamentals and application with PIC. CRC Press, Boca Raton
5. Harper A (2000) High performance printed circuit boards. McGraw-Hill, New York
6. Thierauf SC (2004) High-speed circuit board signal integrity. Artech, Boston

Chapter 6
Applications

Abstract In Microelectronics examples of technological solutions can be considered. In particular Complementary Metal Oxide Semiconductor sensor, Insulated Gate Bipolar Transistors and data converter are the most elements used for the realization of electronics devices. The goal of this last paragraph is to analyze the main features with circuit layouts of some active devices.

6.1 Data Converter

The conversion of analog data into a digital format implies a quantization of the signal with subsequent introduction of a small amount of error. The result is a sequence of digital values suitably stored and/or transmitted at high speed. In according with the applications, the technique of the technique of A/D conversion (ADC) must to be chosen and provides the best compromise between sampling rate and resolution. The process of opposite nature, the digital/analog conversion (DAC), has its wide use in the digital controls of TV and recording sounds. Many ways have been developed to convert an analog signal. The choice of the ADC for a given application is usually defined by design requirements that relate to the choice of a faster, more precise, and compact solution. The output of a microphone, the voltage of a photodiode, or the signal of an accelerometer are examples of analog values which must be converted into digital values so that a microprocessor can handle them appropriately. The successive approximation converter (ADC) evaluates a bit at a time, from the most significant to the least significant bits, using the DAC and comparing the sampled signal with the input signal in feedback. This converter is limited only by the requirements of sample-rate and the input noise. Flash converters (Fig. 6.1) have a resistive ladder that divides the reference voltage into 2N equal parts. For each part, a comparator compares the input signal with the voltage supplied from that of the resistive ladder. A dedicated component called "Priority Encoder" translates the comparison in a binary code. Finally, another type of A/D converter is the sigma-delta that samples the signal in a frequency much higher than the Nyquist frequency. For this reason it is also called oversampling converter. As regards the process of digital/analog conversion, the most common type of converter is represented by the pulse width modulator (PWM). A stable voltage or current is switched to a low-pass analog filter with a duration determined by the digital input code. This technique is often used to control the speed of the electric motor and

M. Di Paolo Emilio, *Microelectronics*, DOI 10.1007/978-3-319-22545-6_6

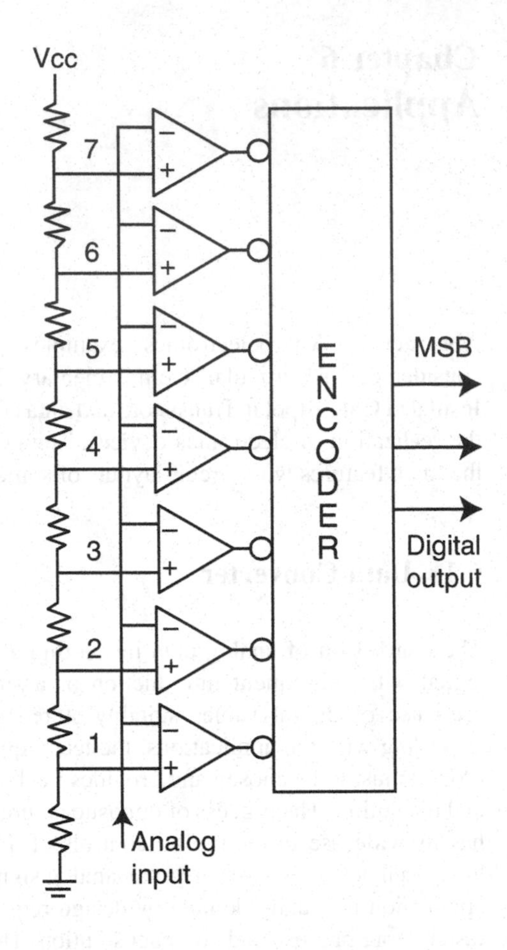

Fig. 6.1 ADC flash

in many other applications. Other types of D/A converters are the DAC resistors weighings, which contains individual electrical components for each bit of the DAC connected to a summing point; and the R–2R, a binary weighted DAC (Fig. 6.2) which uses a cascade repeated structure of resistance with values R and 2R. The first parameter to consider in a D/A is the Full Scale, FS, or the maximum voltage capable of converting which is linked to the resolution of the entire system. Further parameter is the maximum sampling frequency that expresses the speed at which the DAC circuitry can operate and produce the correct output. All ADCs suffer from nonlinearity errors caused by their physical imperfections and the corresponding resolution that indicates the number of discrete values produced in the range of analog signal. For a simplicity of cost, the signals are often sampled at the minimum rate required, with the result that the quantization noise introduced is white noise and is spread over the entire bandwidth of the converter. If it is sampled at frequencies that do not fulfill the requirement of Nyquist, instead, they are being falsely detected samples of noise in a process called Aliasing [1, 2].

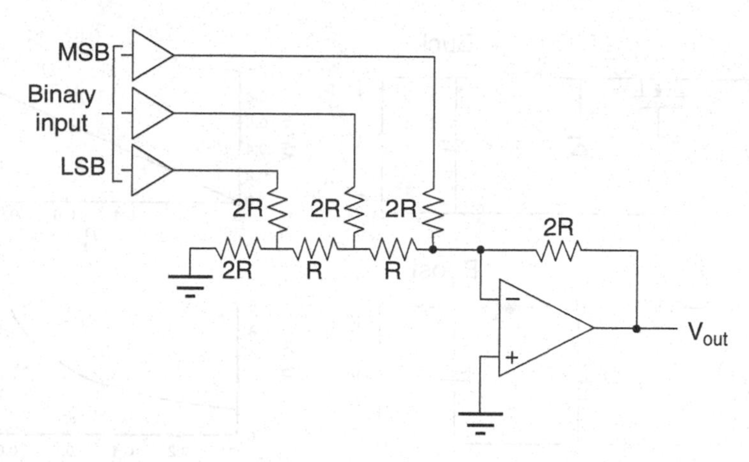

Fig. 6.2 DAC R–2R

6.2 DC–DC Converter

A DC–DC converter is an electronic circuit with outputs DC voltage at a different level compared to the input (DC). They are used in consumer devices such as mobile phones and tablet with the goal of providing isolation from the noise and a power control due to the fact that many electronic systems contain several sub-circuits which operate at different voltage levels, thereby it is possible to save space and costs. The DC–DC power converters are employed in a variety of applications, including power supplies for personal computers, office equipment, power systems of space vehicles, portable computers and telecommunications equipment, as well as motor drive current. In particular, the DC–DC converter reduces the battery voltage from 5 to 3.3 V required by many integrated circuits of the processor and has an ideal efficiency of 100 %; in practice, efficiencies from 70 to 95 % are typically commercial values. The efficiency can be increased by using another active components as MOSFET instead of bipolar transistors or Shottky diodes.The pulse width modulation (PWM) allows the control and regulation of the output voltage. This approach is also used in applications involving AC power converters including DC–AC (inverter and power amplifiers), AC–AC, and some AC–DC power converters. The type of conversion can be linear or switching. The linear converters utilize an internal reference and regulate the voltage via a feedback circuit. The converters switches, however, are similar to switching power supplies, where the conversion is done generally by virtue of an inductor that stores the magnetic energy [1, 3, 4]. The switching technology is more efficient than linear and ensures an increase of the battery life for portable devices. One disadvantage of switching converters is the electrical noise generated at high frequencies that can, however, be limited using suitable filters. In Fig. 6.3 are shown various circuits of DC–DC converters commonly used, together with their respective conversion

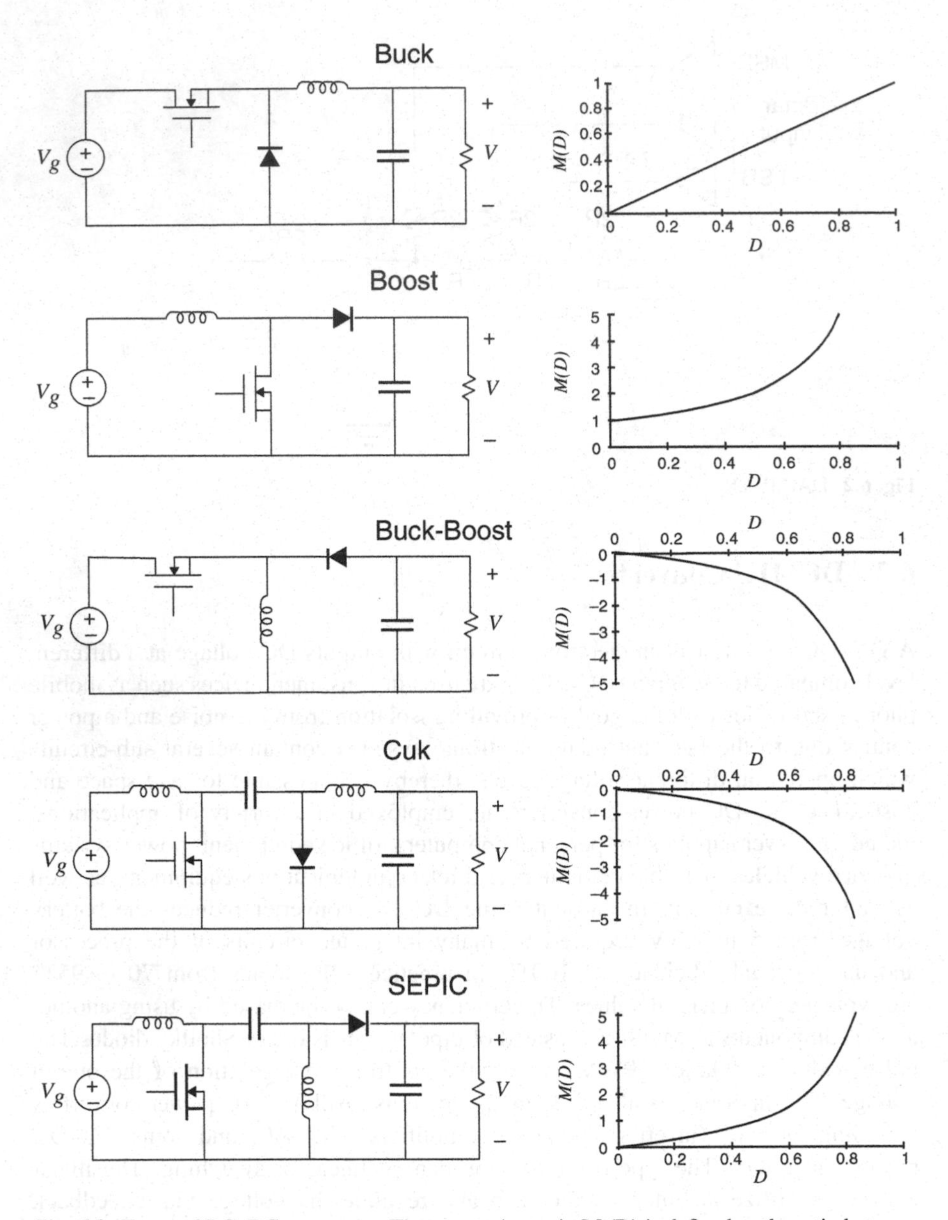

Fig. 6.3 Types of DC–DC converters. The conversion ratio M (D) is defined as the ratio between the output voltage (Vcc) and the input voltage (Vg) in stationary conditions; D is the duty cycle

ratios. In each example, the switch is made by using a power MOSFET and diodes; however, other semiconductor switches such as Insulated Gate Bipolar Transistors (IGBTs) may be used.

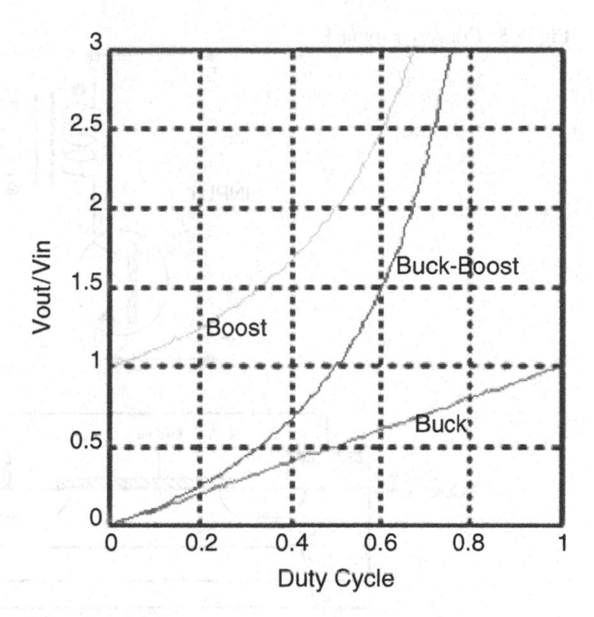

Fig. 6.4 Ratio V_{out}/V_{in} as a function of the duty cycle

The buck converter reduces the voltage and the conversion ratio is M (D) = D. In a similar topology, known as a boost converter, the inductor and switch positions are interchangeable. This converter produces an output voltage V that is greater of the input voltage V_g. The conversion ratio is $M(D) = 1/(1 - D)$. In the buck–boost converter, the switch alternately connects the inductor through the input and output voltages. This converter inverts the polarity of the voltage and can increase or decrease the amplitude (Fig. 6.4) of output; the conversion ratio is $M(D) = -D/(1 - D)$ [3].

The Cuk converter, instead, is essentially a boost converter followed by a buck converter with a capacitor to couple the energy. The network of switches alternately connect a capacitor to the input and output inductor. The conversion ratio $M(D)$ is identical to that of the buck–boost converter. The single-ended primary inductance converter (SEPIC) can also increase or decrease the tension. However, it does not reverse the polarity. The conversion ratio is $M(D) = D/(1 - D)$. In many DC–DC applications, multiple outputs and the input/output insulation are required. These characteristics meet safety standards and provide correspondences of impedances. DC–DC topologies described above can be adapted to these features; an example is the flyback converter that can be developed as an extension of the Buck–Boost converter (Fig. 6.5).

6.3 Insulated Gate Bipolar Transistor

The IGBT are semiconductor devices with high input impedance, capable of switching high voltages and currents. Many designers see the IGBT as a device with input characteristics of MOS and bipolar output characteristic that is a

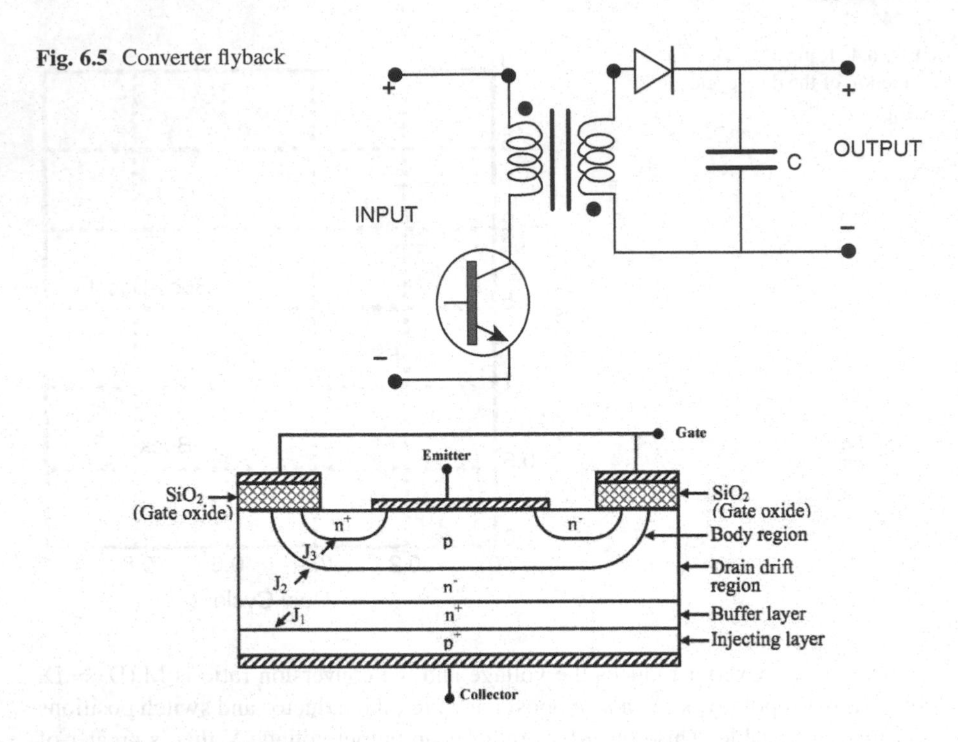

Fig. 6.5 Converter flyback

Fig. 6.6 Layout of an IGBT

voltage-controlled bipolar device. It is essentially a functional integration of power devices MOSFET and BJT in monolithic form, used in many applications of power electronics, especially in three-phase PWM drives that require a high dynamic range control and low noise. Moreover, it can be used in uninterruptible UPS and Switched-Mode Power Supplies (SMPS). The basic schematic of a typical IGBT N-channel DMOS is shown in Fig. 6.6. It is evident that the cross section of an IGBT is almost identical to that of a power MOSFET, exception made for the P+ layer. The N+ layer at the top is the source or emitter and P+, at the bottom, is the collector or drain [5]. Several advantages can be analyzed with respect to a MOSFET and BJT, in particular the very low on-state voltage makes the device with smaller dimensions and low cost. It presents a simple driving circuit easily controlled in high voltage and located in space applications of high current. It can manage current higher than bipolar transistors with switching speed less than of a power Mosfet but more than a BJT. In case of a power MOSFET, the resistance-on increases strongly with the breakdown voltage due to an increase in resistivity and thickness of the drift region to support the high operating voltage. For this reason, the development of high-power MOSFET with high blocking voltage is normally avoided. In contrast, for the IGBT, the resistance of the drift region is drastically reduced by the high concentration of minority carriers injected during the conduction of current.

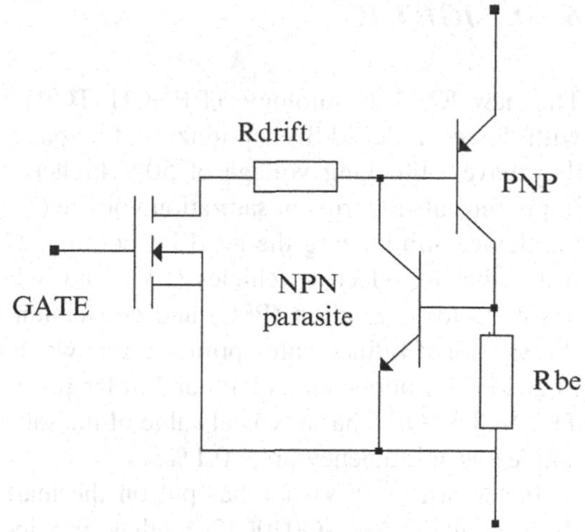

Fig. 6.7 Equivalent circuit of an IGBT

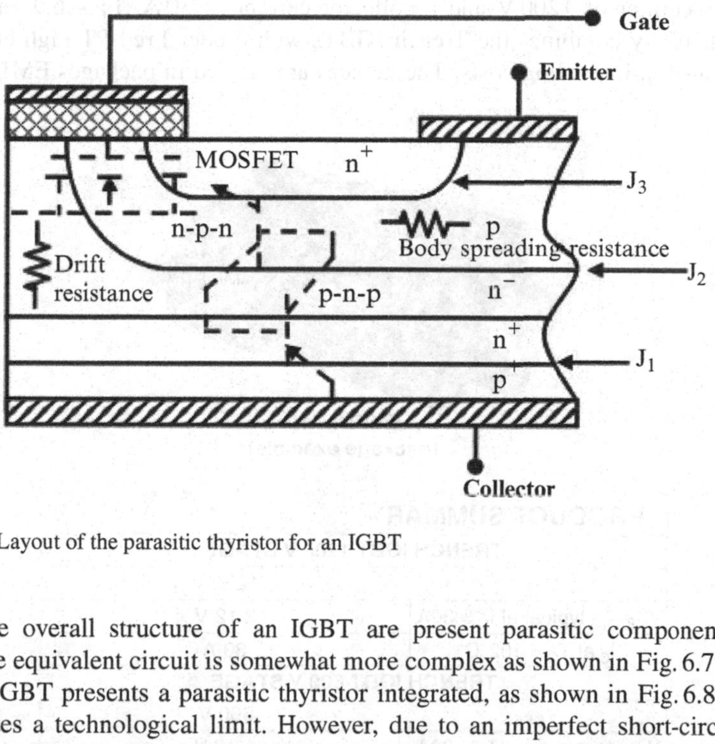

Fig. 6.8 Layout of the parasitic thyristor for an IGBT

In the overall structure of an IGBT are present parasitic components. The complete equivalent circuit is somewhat more complex as shown in Fig. 6.7.

The IGBT presents a parasitic thyristor integrated, as shown in Fig. 6.8, which constitutes a technological limit. However, due to an imperfect short-circuit, the exact equivalent circuit of the IGBT includes the resistance between the base and the emitter of the npn transistor. If the output current is large enough, the voltage drop across the resistor may jeopardize the polarization of the npn transistor and a malfunction condition which can destroy the device [5–8]. It's very important that the parasitic thyristor does not trigger during device operation to prevent an uncontrolled flow of current which can destroy the device itself.

6.3.1 *IGBT IC*

The new IGBT technology TRENCHSTOP5 Infineon is based on thin wafers, with losses reduced by optimizing the package. Presented in recent months, they have a blocking voltage of 50 V higher than the previous versions. Further improvements in terms of saturation voltage (V_{CE} (sat)) lead to greater reliability of the device, minimizing the need for cooling. Main applications are in automotive fast switching, electric vehicles (EV), and hybrid (HEV) in applications such as power factor correction (PFC) and conversion DC/AC and DC/DC. More IGBT (Series L5) of Infineon are optimized for switching frequencies ranging from 50 Hz to 20 kHz for applications UPS and Solar Inverters. The series L5 with technology TRENCHSTOP 5 has a typical value of the saturation voltage V_{ce} of 5.1 V at 25 ° C and levels of efficiency up to 0.1 %.

In the last year, Vishay has put on the market a series of IGBT, for example, the module VS-ENQ030L120S that provides a breakdown voltage of the collector–emitter of 1200 V and a collector current of 30 A (Figs. 6.9 and 6.10). The technology combines the Trench IGBTs with diodes Fred PT high efficiency for a reduction in cooling costs. The devices are offered in packages EMIPAK-1B

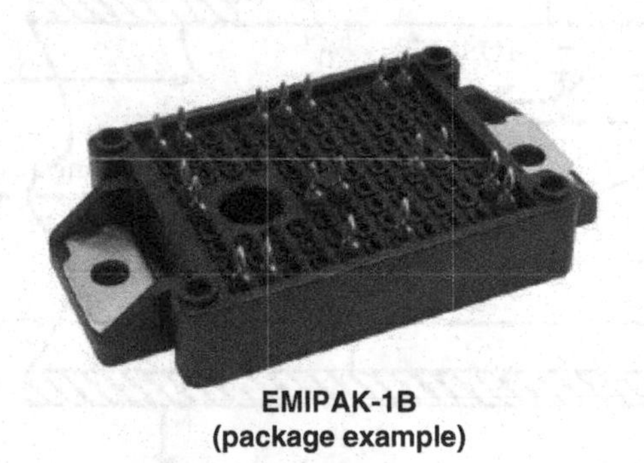

PRODUCT SUMMARY	
TRENCH IGBT 1200 V STAGE	
V_{CES}	1200 V
$V_{CE(ON)}$ typical at I_C = 30A	2.12 V
I_C at T_C = 102°C	30 A
TRENCH IGBT 600 V STAGE	
V_{CES}	600 V
$V_{CE(ON)}$ typical at I_C = 30A	1.42 V
I_C at T_C = 106°C	30 A
Package	EMIPAK-1B
Circuit	3-Levels Neutral Point Clamp Topology

Fig. 6.9 VS-ENQ030L120S

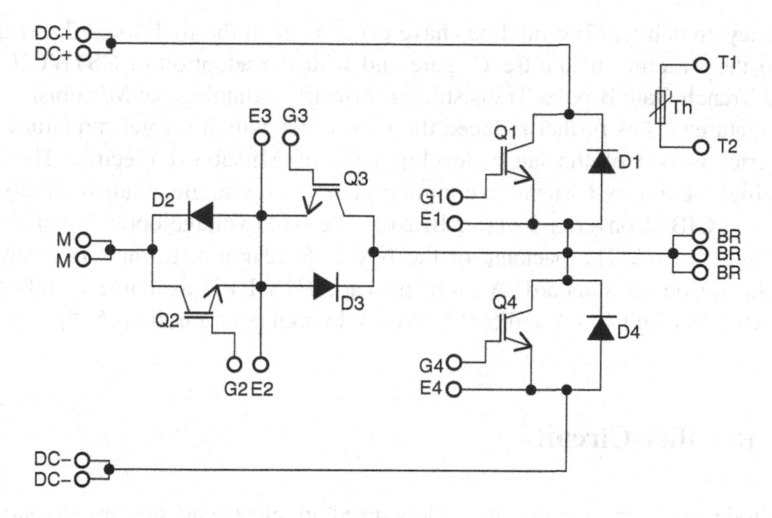

Fig. 6.10 Configuration mode of VS-ENQ030L120S

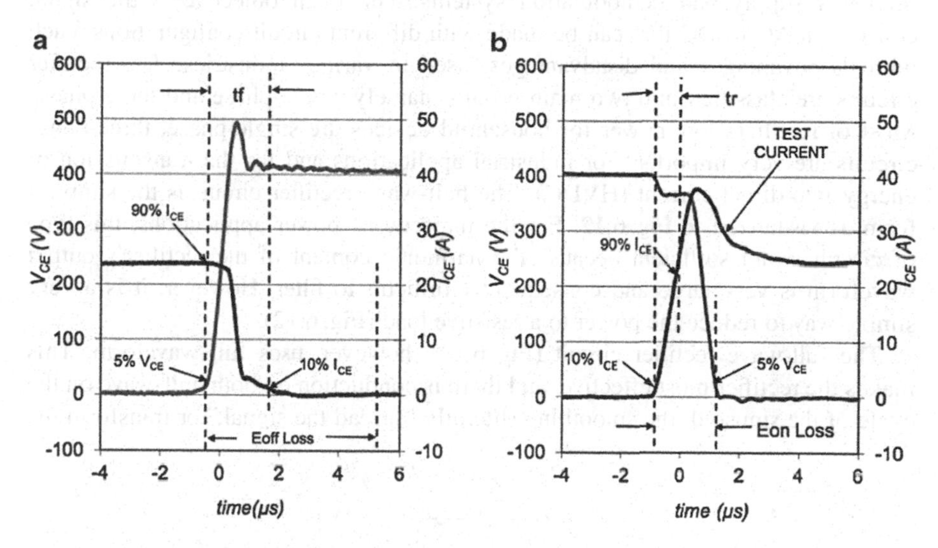

Fig. 6.11 Feature Vce for AUIRGP66524D0 (**a**) and AUIRGF66524D0 (**b**)

(VS-ENQ030L120S) and EMIPAK-2B (VS-ETF075Y60U, VS and VS-ETF150Y65U-ETL015Y120H). The International Rectifier (IR) has recently entered in the market for the IGBT 600 V for automotive: AUIRGP66524D0 and AUIRGF66524D0 with TO-247 package (Fig. 6.11) used in electric and hybrid vehicles. The technology of IR called COOLiRIGBT 24 A is characterized by a low value of V_{CE} (ON) of about 1.6 V ensuring low power consumption. The line of products of Mitsubishi includes various types of IGBT modules, such as the NF series that has retained its traditional form and the series NFH for the high frequency switching. The modules have evolved from the traditional flat structure toward the structure of the trench gate and with the adoption of CSTBT (Carrier Stored Trench Gate Bipolar Transistor, proprietary technology of Mitsubishi). With these features it has further reduced the power loss with a greater miniaturization. NX-Series is one of the latest developments of Mitsubishi Electric. The series offers high flexibility by using a common platform for single, dual, six-and-seven-packs and CIB (Converter-Inverter-Brake). The wide voltage options include 600, 1200, and 1700 V. The package of the NX-M (122 mm × 62 mm) is compatible with the European standard. A large package (NX-L) is available in 600 A and 1000 A/1200 V and 400 A and 600 A/1700 V in dual configuration [5–8].

6.4 Rectifier Circuits

The diode rectifier circuits are widely used in electronic design, in particular in power supply and demodulation systems. The main objective is the signal conversion AC to DC that can be made with different circuit configurations, each with its advantages and disadvantages, used in various industries. The rectifier circuits are classified into two main groups, namely single-phase and three-phase. Most of rectifiers low power for household devices are single-phase, three-phase circuits are very important for industrial applications and for the transmission of energy into direct current (HVDC). The half-wave rectifier circuit is the simplest form as visualized in Fig. 6.12. For the majority of power applications, this type of circuit is not sufficient because the harmonic content of the rectifier's output waveform is very large and consequently difficult to filter. However, it is a very simple way to reduce the power to a resistive load (Fig. 6.12).

The full-wave rectifier circuit (Fig. 6.13), however, uses full waveform. This makes the rectifier most effective, and there is conduction on both half-wave on the cycle of the sinusoid, the smoothing (literally "spread the signal," or transform the

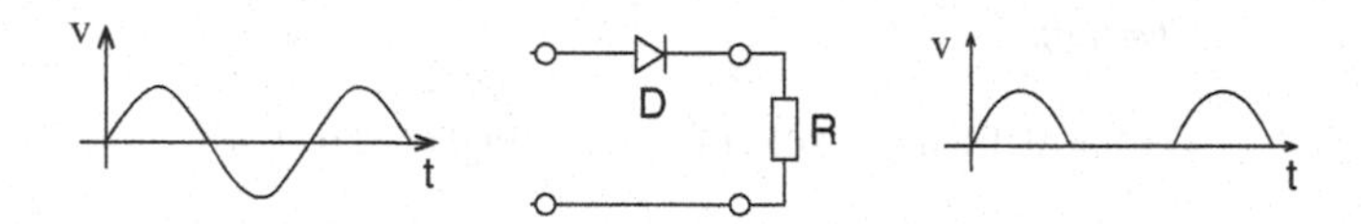

Fig. 6.12 General layout of a half-wave rectifier

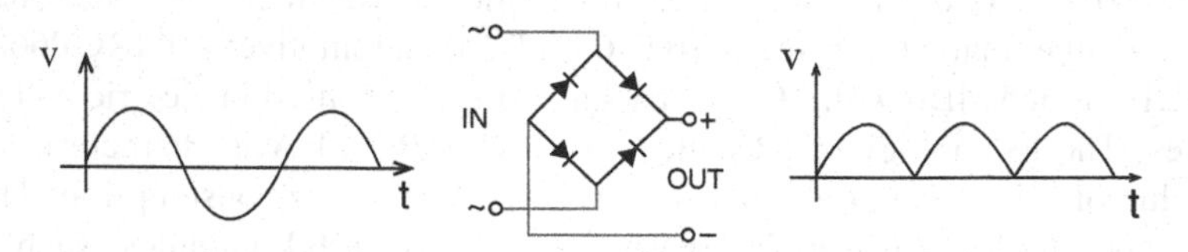

Fig. 6.13 General layout of a full-wave rectifier

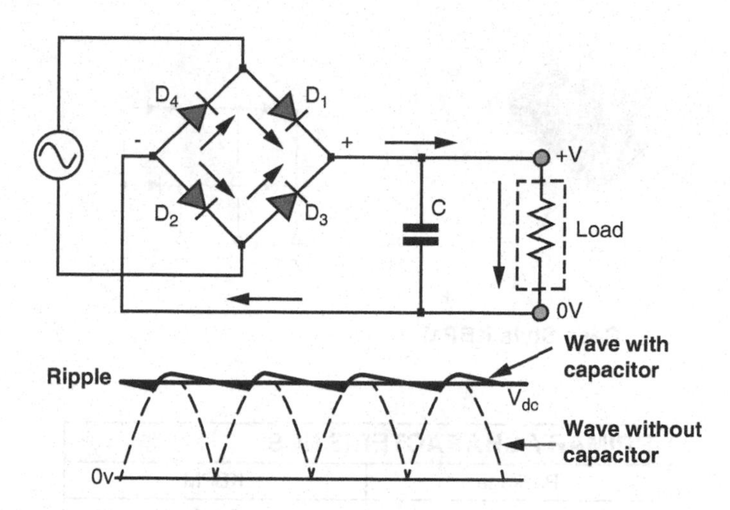

Fig. 6.14 Example of bridge rectifier circuit

AC signal into DC) becomes much easier and more effective. The classical circuit is based on four diodes in a bridge topology. The diodes can be replaced with active elements to provide the switching and increase the efficiency. The choice depends on the application of the circuit. While the full-wave circuits are mostly used with the bridge configuration, half-wave circuits may offer a better solution in some circumstances. For power applications are normally used power Schottky diodes which require only a forward voltage of about 0.2–0.3 V.

6.4.1 Bridge Rectifier Circuit

The bridge rectifier (single phase) uses four rectifier diodes, each connected in a closed-ring configuration ("bridge") to produce the desired output. The main advantage of this bridge circuit is that it requires a dual transformer. The diodes (D1 to D4 in Fig. 6.14) are arranged in series, only two diodes conduct current during each half cycle of the sinusoidal waveform. During the positive half cycle, diodes D1 and D2 are in conduction, while the diodes D3 and D4 are reverse biased and current flows through the load as shown in Fig. 6.14. Conversely, during the negative half cycle [5–8].

Bridge rectifiers integrated components are available in a range of different voltages and dimensions that can be welded directly into a PCB circuit (Fig. 6.15). Depending on the technology of the diode threshold voltage, each may vary in the neighborhood of 0.6 V.

For a complete description, several parameters must be considered: for example, repetitive peak voltage (VRRM) and the Reverse Recovery Time. The repetitive peak voltage (VRRM) is the actual value of maximum allowable reverse voltage across the diode rectifier.

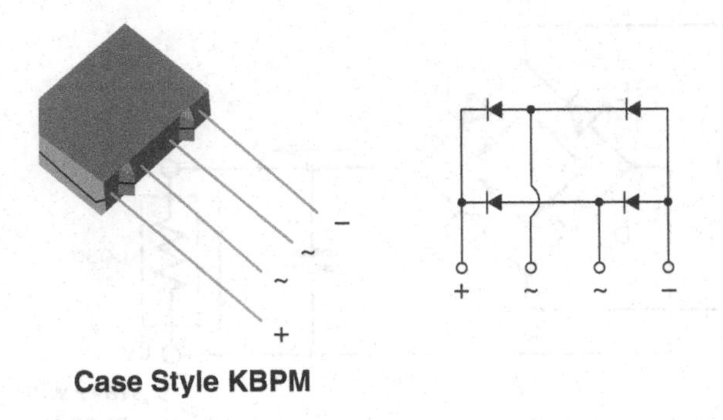

PRIMARY CHARACTERISTICS	
Package	KBPM
$I_{F(AV)}$	2.0 A
V_{RRM}	50 V to 1000V
I_{FSM}	60 A
I_R	5 μA
V_F	1.1 V
$T_J$max.	165 °C
Diode variations	In-Line

Fig. 6.15 Example of diode bridge of Vishay

When switching from the conducting to the blocking state, a rectifier has stored charge that must first be discharged before the diode blocks reverse current. This discharge takes a finite amount of time known as the Reverse Recovery Time (t_{rr}).

The smoothing capacitor placed in the output (Fig. 6.14) converts the wave into a DC voltage. Generally, in the DC power circuits, the smoothing capacitor is an electrolytic that has a capacitance value of the order of 100 uF or more. However, two parameters are important to consider when choosing a smoothing capacitor, those are its operating voltage, which must be greater than the value of load output of the rectifier, and its capacitance value, which determines the amount of ripple (variable part of the continuous signal) which will be superimposed on the DC voltage. As a general rule, you can think of a ripple voltage of less than 100 mV peak to peak. The amount of ripple voltage can be virtually eliminated by adding a filter to the output terminals of the rectifier bridge. This low-pass filter is constituted by two capacitors, generally of the same value, and an inductance in order to induce a path of high impedance for the alternating ripple component. Another alternative layout most practical and economical is to use a voltage regulator, such as an LM78xx (where "xx" stands for the output voltage) that can reduce the ripple of over 70 dB offering at the same time a constant output current of more than 1 A [7, 8].

6.4.2 Design Considerations

An important aspect of the rectifier circuits is the loss in the output voltage, caused by the voltage drop of the diodes (about 0.6 V for silicon and 0.3 V for the Schottky diodes). This reduces the output voltage limiting, therefore, that available output. The loss of voltage is very important for low voltage rectifiers (for example, 12 V or less), but is insignificant in high voltage applications such as HVDC. The rectifiers are also used for the detection of the amplitude modulated signal relatively to radio signals. Another typical use of rectifier circuits is the power supply design. A power supply can be divided into a series of blocks (Figs. 6.16 and 6.17), each of which performs a particular function: a transformer and a rectifier circuit for converting the AC signal into DC. The power supplies are designed to produce less ripple that can cause several problems. For example, in audio amplifiers, too much ripple looks like an annoying buzzing noise; in video circuitry, an excessive ripple causes defects in the image; in digital circuits can cause erroneous results of logic circuits.

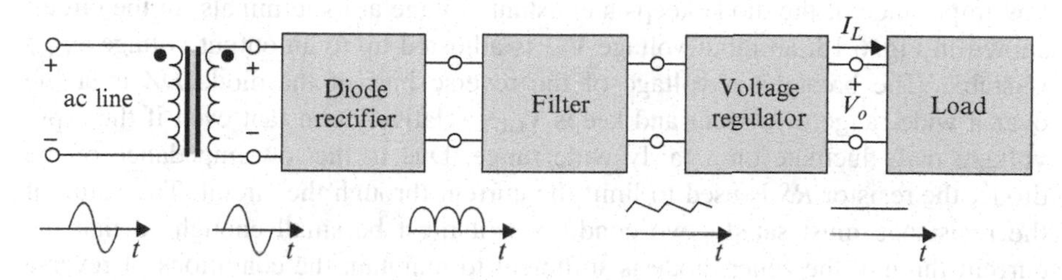

Fig. 6.16 Block diagram of a power supply

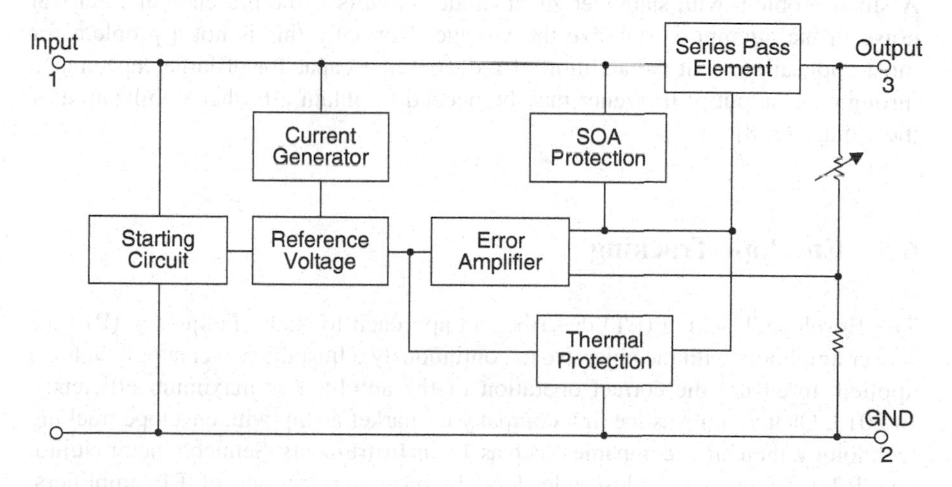

Fig. 6.17 General diagram of power supply

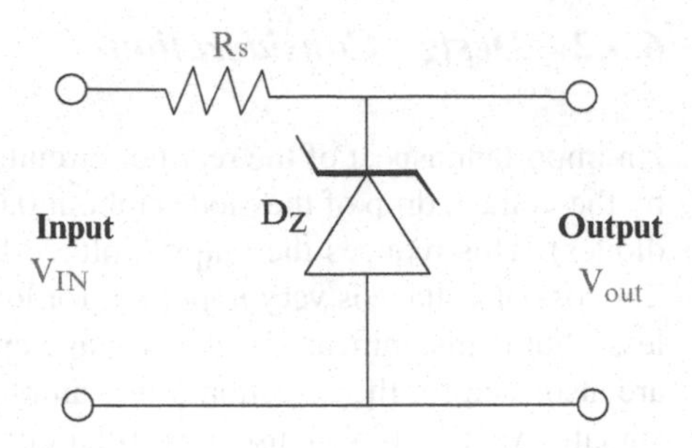

Fig. 6.18 General diagram of a voltage regulator with zener diode

6.4.3 *Zener Diode as Voltage Regulator*

The Zener diodes are widely used in reverse bias to produce a stabilized output voltage. When connected in parallel to a variable voltage generator in reverse bias, such as the rectifier diode circuits just discussed, the zener diode conducts when the voltage reaches its reverse breakdown voltage. From that moment on, the relative low impedance of the diode keeps a constant voltage at its terminals. In the circuit shown in Fig. 6.18, an input voltage V_{IN} is adjusted up to an output voltage V_{OUT} is stable. The breakdown voltage of the reverse bias of the diode *DZ* is stable over a wide range of current and keeps V_{OUT} relatively constant even if the input voltage may fluctuate on a fairly wide range. Due to the low impedance of the diode, the resistor *RS* is used to limit the current through the circuit. The value of the resistance must satisfy two conditions: it must be small enough so that the current through the zener diode is sufficient to maintain the conditions of reverse breakdown; moreover, It must also be large enough for the current through *DZ*. A small problem with stabilizer zener diodes circuits is the presence of electrical noise in the attempt to stabilize the voltage. Normally this is not a problem for most applications, but the addition of a decoupling capacitor of large capacitance through the output of the zener may be needed to obtain a further stabilization of the voltage [7, 8].

6.5 Envelope Tracking

The Envelope Tracking (ET) describes an approach to Radio Frequency (RF) for power amplifiers with the objective to continuously adjust the power supply voltage applied, to ensure the correct operation of the amplifier at maximum efficiency. In 2013, Qualcomm was the first company to market a chip with envelope tracking technology, then other companies such as Texas Instruments, Semiconductor Nujira, and R2 have become industry leaders. In most applications of RF amplifiers, efficiency has a significant impact on the design. In an RF transmitter, most of the power is used by the power amplifiers (PA): typically, more than 50 % of the input power is dissipated. Therefore, improving efficiency in the RF power amplifier is

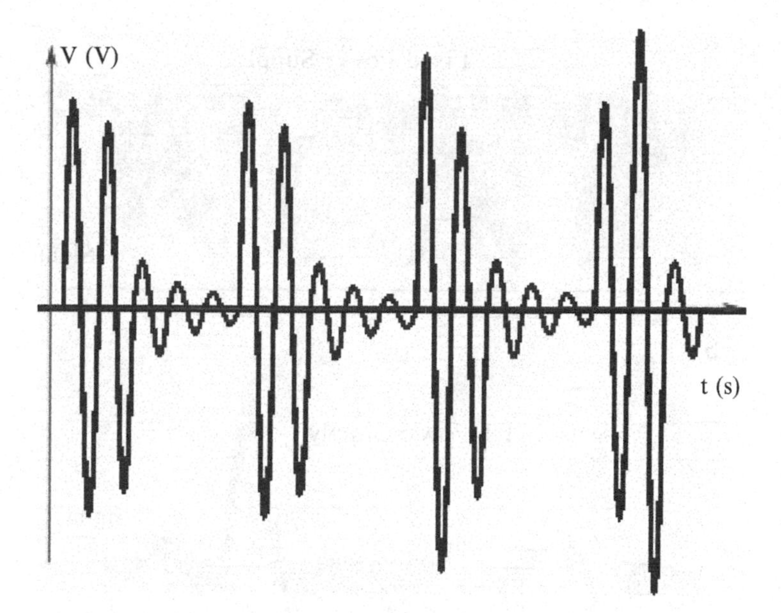

Fig. 6.19 Example of signal peaks with high energy

a crux of numerous research activities and the Envelope Tracking (ET) is a great option. Other techniques used are the Crest Factor Reduction (CFR) and Digital Pre-Distortion (DPD). The efficiency of an ET is improved by varying the supply voltage of an amplifier in synchronism with the envelope of the RF signal. The fundamental characteristics of output of an RF power amplifier, such as the power, efficiency, gain and phase, depend on two control: RF input power and power supply voltage. The efficiency of an amplifier depends also on the waveform and the operating mode. For data transmission systems used today such as UMTS, HSPA, and LTE 4G, the RF waveforms are used to incorporate an amplitude in addition to phase elements, and therefore, require a linear amplifier.

During the signal peaks as shown in Fig. 6.19, the amplifier requires the full voltage in order to provide the necessary power without incurring in errors, but during periods with lower levels is not required this tension and a certain power is dissipated in the device. The amplifier only requires a lower voltage to provide the lowest levels of power and, therefore, working with the highest voltage all the time, there is a useless waste of energy. Envelope Tracking is a technique that requires a supply voltage of the final stage (PA) dynamically modulated with the envelope of the input signal. In this case the PA may operate, at any time, closer to the peak level and dramatically the efficiency of the system is improved by reducing the amount of energy dissipated as heat (Fig. 6.20). QFE1100 of Qualcomm was the first ET system for a mobile 4G (Samsung Galaxi 3).

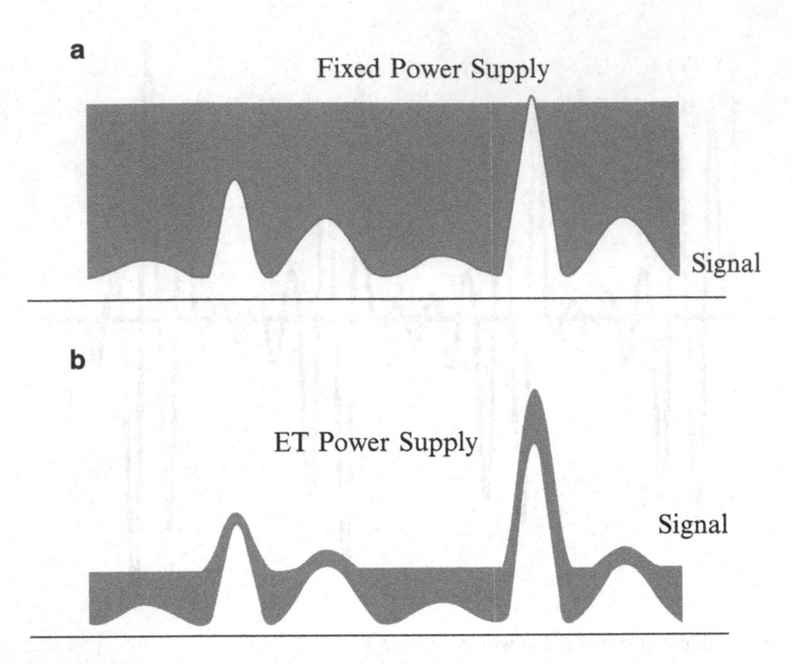

Fig. 6.20 Comparison of the traditional technique of power supply (**a**) and the ET system (**b**)

6.5.1 Block Diagram of the Traditional System

One traditional scheme of RF amplifier is shown in Fig. 6.21 and consists of a system where the modulated signal is created and converted to the final frequency. Subsequently, it is amplified and applied to the final RF power amplifier. Traditionally it is synchronized by a DC–DC converter which provides a constant voltage.

Most of the current configurations are in a digital format as displayed in Fig. 6.22.

In the diagram it is observed that the signals I and Q in their digital format are separately applied to a digital-analog converter to convert them into an analog format. The signal is sent, subsequently, in a low pass filter to remove the alias and unwanted higher frequency products. The signals, then, are mixed with the local oscillator to bring them to the desired frequency, they are then added together to create the final signal, and then sent to the amplification chain to obtain the amplitude level required [7, 8].

6.5.2 Envelope Tracking system

Adding a DC–DC converter to the traditional amplification configuration, a Envelope tracking system (Fig. 6.23) is defined in order to provide the correct voltage output to the power amplifier so that it can dissipate the minimum amount of energy.

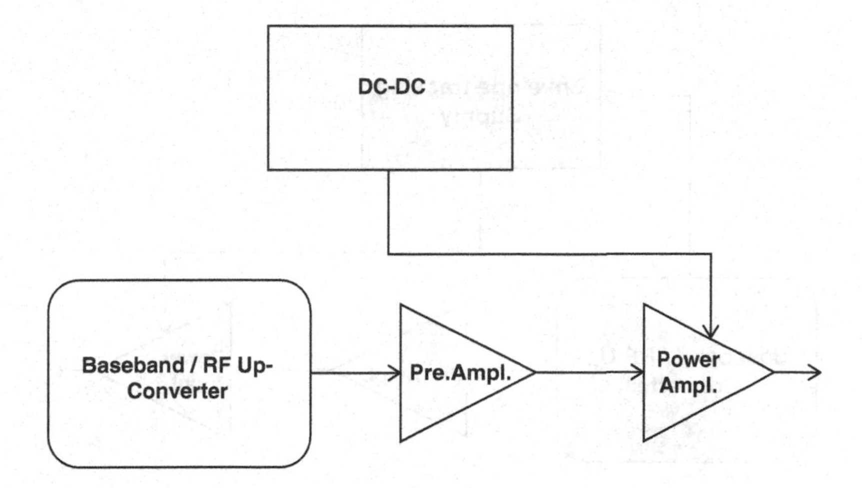

Fig. 6.21 Block diagram of a conventional RF amplifier

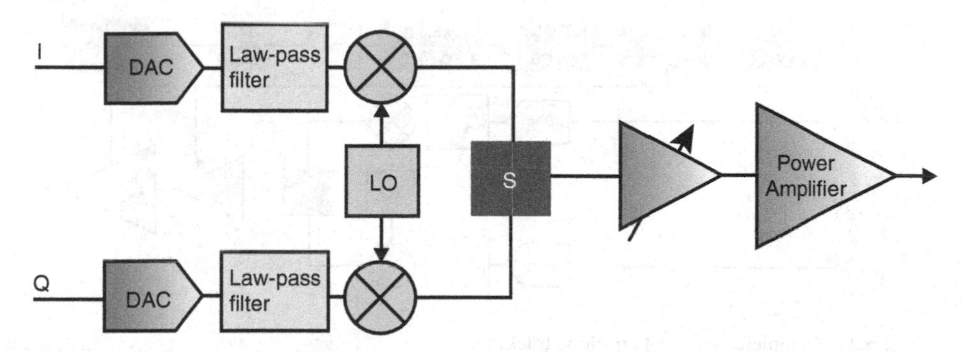

Fig. 6.22 Block diagram of a conventional amplifier in the digital configuration

The system of Envelope Tracking incorporates a new series of blocks as represented in Fig. 6.24. These can be summarized below:

- Lock RF: The I and Q signals are used to create the composite signal RF that is sent to the RF amplifiers.
- Envelope Tracking Supply: it modulates the voltage of the power amplifier providing the maximum efficiency. The chipset from Texas Instruments, LM3290 and LM3291 enable significant energy savings for Envelope tracking RF systems. In particular, LM3290 and LM3291 work with integrated DC–DC boost converter optimized for technique of tracking in RF power amplifiers (PA). The device allows the maximum transmit output power independent of the input voltage of the battery (battery from 2.5 V) and is controlled by the MIPI RFFE 1.1. In addition, the family of the IC Nujira Coolteq are employed in the field of 3G and 4G smartphones. With a bandwidth of 20 MHz, the NCT-L1300 offers energy savings by using the industry standard MIPI RFFE eTRAK and interfaces (Fig. 6.25).

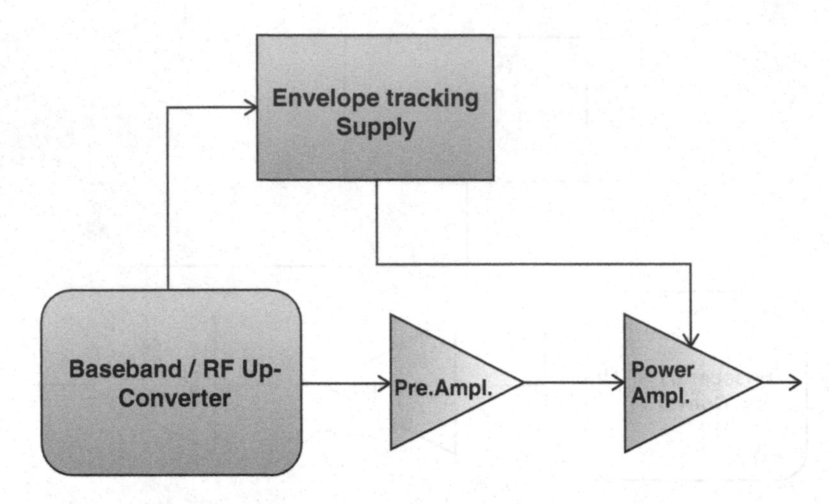

Fig. 6.23 Block diagram of an envelope tracking system

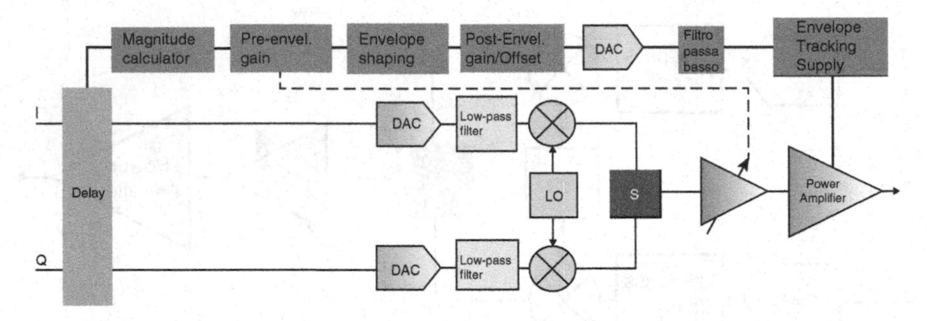

Fig. 6.24 Complete system of envelope tracking

- Envelope Shaping: generates the envelope signal required.
- Delay: various delays to ensure synchronism between RF and envelope.

Some of the requirements for the Envelope Supply systems are the following:

- Bandwidth: must be able to accurately follow the modulation envelope at higher frequencies. To follow with precision the modulation must be able to reach frequencies typically about two-three times that of the signal band. For the current systems, this may require levels of bandwidth of 50 MHz.
- Noise: although switching power supplies offer the highest levels of efficiency, one of the challenges of their use is the switching noise. An innovative design is necessary to ensure high efficiency in terms of noise.
- No decoupling capacitor: it is a standard practice in conventional power supplies add a decoupling capacitor in output to avoid the noise and ripple. Considering the high bandwidth required, the power supply may not have any decoupling capacitor in output.

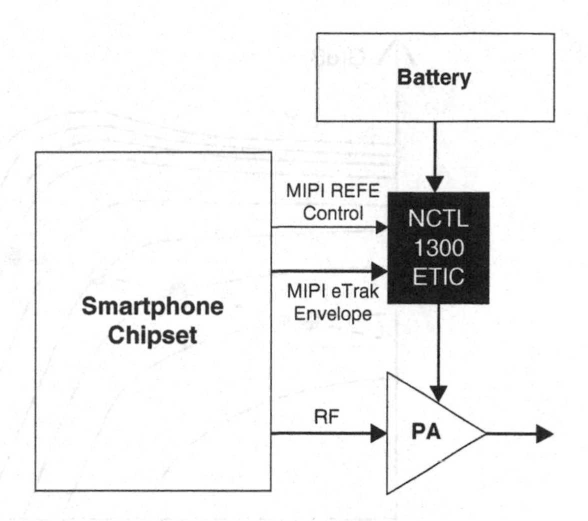

Fig. 6.25 NCT-L1300 of Nujira Coolteq

- Low output impedance: considering the fact that no decoupling capacitor is allowed, ET systems must have a low output impedance that extends up to the maximum frequency of modulation.

Moreover, it is particularly important to have a good synchronization between the RF Envelope and the control signal, otherwise important consequences deteriorate the performance of RF amplifier, in particular:

- Increase of the power dissipation: if there is an insufficient synchronization between the waveforms, the power dissipation increases because the voltage peaks are not synchronized with the peaks of the RF envelope.
- Efficiency is reduced: the power dissipation in electronic components, is reflected in the overall operation of the amplifier with reduced of the efficiency (Fig. 6.26).
- Distortion is increased: insufficient synchronization means that there will be insufficient power to envelope tracking and this will also cause a distortion in terms of signal.

6.5.3 Test Solutions

The synchronization is a crucial specification of a test configuration for ET systems. By reason of the stringent requirements of synchronization, the PXI platform is well suited to address the challenges of testing with modular instruments interconnected through a chassis backplane containing a number of clock lines and distribution. This single chassis simplifies the configuration of the equipment and improves system synchronization. In addition to the advanced PXI hardware, National Instruments offers the LabVIEW programming environment with the ability to process and display signals in real time, improving the development

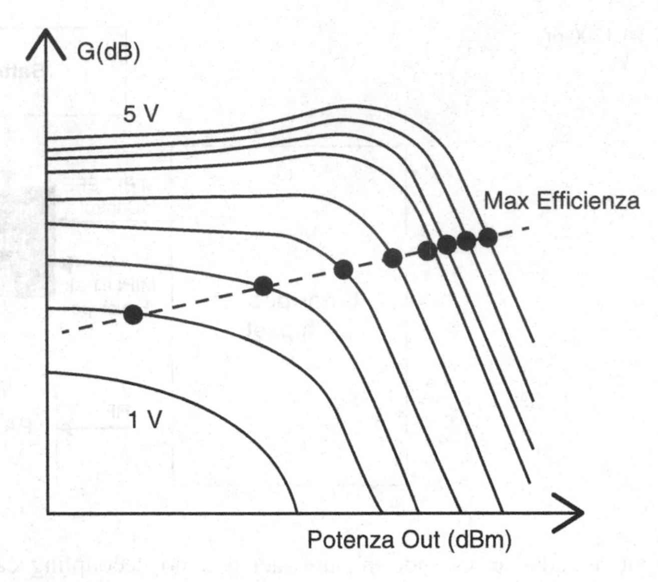

Fig. 6.26 Maximum efficiency of the ET system

of devices for this application. An example is the NI PXIe-5451 AWG and the vector signal transceiver NI PXIe-5644R with the maximum sync jitter less than 50 ps. Rohde & Schwarz also offers a measure of all-in-one to meet the growing needs of chipset manufacturers for characterizing power amplifiers with solutions of tracking. This solution includes the signal generator R & S-SMW200A with the use of any communication standards, including LTE or WLAN and the spectrum analyzer R & D-FSW. Users can delay the RF signal and the envelope signal relative to each other in real time from $\pm$ 500 ns with a resolution of 1 ps. This guarantees a perfect synchronization between the supply voltage and the modulated RF signal. In reason of Envelope Tracking amplifiers operate normally in the linear range, the distortion is proven to be important for defining the efficiency of the system. Considering the deterioration of the RF performance, it is often used the R & D-SMW-K541 option of digital predistortion. The Keysight offers the option for Envelope Tracking Signal Studio LTE/LTE-Advanced (N7624B/25B-KFP) that provides an easy way to generate I and Q signals (RF). The function of the Signal Tracking Control Studio allows the alignment of RF signals and envelope, with a resolution of 1 ps without the need to regenerate the waveforms. This allows fine adjustment of the timing by monitoring parameters Adjacent Channel Power (ACP) or Error Vector Magnitude (EVM). The parameter ACP measures the ratio between the total power adjacent channel (intermodulation signal) to the main channel's power (useful signal). Mobile communication systems have relied on measurements ACP to ensure a good signal to noise ratio (SNR) in the adjacent channel to avoid interference in communications. EVM, instead, is a measure used to quantify the performance of a digital radio transmitter or receiver. It is a measure of the modulation quality and error performance in complex wireless systems. In recent years Envelope Tracking systems have tried to be an efficient tool to increase the efficiency of the power amplifier, reducing the cooling requirements. While systems ET promise a considerable energy saving and longer battery life than conventional power supplies, new challenges for designers and test engineers become necessary for a good design both in terms of efficiency and in terms of cost and time-to-market.

6.6 CMOS Sensors

The sensors of digital cameras are based on the technology of the CCD or Complementary Metal Oxide Semiconductor (CMOS). Both have similar properties and work on the physical principle to convert light into electric charges. The CMOS sensors are cheaper and using the production technique of microprocessors (CPU) and memory chips. A digital sensor is an array of pixels (matrix); each is a receptor of photons. The processor, inside the camera, calculates the number of photons captured by each pixel. The result is translated into bit depth (from 0 to 255 for 8-bit images). To evaluate also the color, a Bayer filter is applied to each pixel to select the corresponding wavelengths.

The matrix is formed by a set of rows red, green, and blue. Since the human eye is more sensitive to green color and to reduce digital noise, the number of pixels of green color appears to be double (50 %) with respect to the other two (25 %). Some sensors (Fig. 6.27) do not make use of this Bayer matrix, but interpret the value of each single pixel [7, 8]. The calculation algorithm of the processor named "demosaicing" extracts information from a group of pixels and determines the value in megapixel of the sensor.

Fig. 6.27 CCD (*left*) and CMOS (*right*) sensor

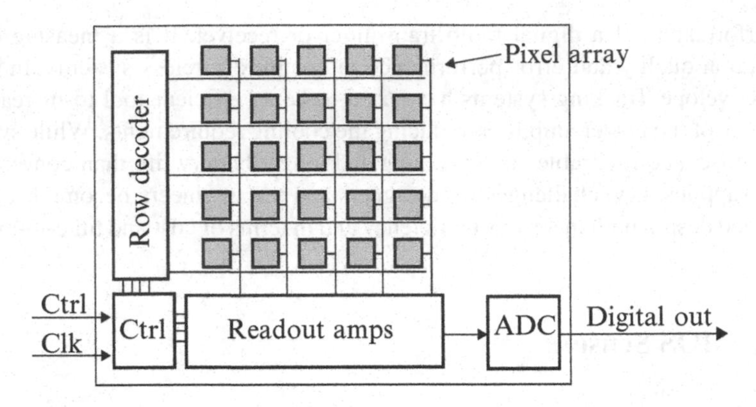

Fig. 6.28 CMOS sensors

6.6.1 CMOS Technology

A CMOS image sensor is a chip that converts the incoming light into electrical signals (analog), and is done in a process called CMOS. The sensor layout is realized to obtain greater reflection to increase the interaction efficiency of photons with the metal layers. If the photon has a higher energy than the gap energy of silicon, which is 1.12 eV, it can excite one of the valence electrons in the conduction band. This is according to the photoelectric effect, for which Albert Einstein received in 1921 the Nobel Prize in Physics. In a CMOS image sensor, there are three different ways of separation and collection of the electron-hole pairs photogenerated: using an array of photodiodes, photogates, or phototransistors. CMOS sensors (Fig. 6.28) have several advantages. Unlike the CCD, the CMOS integrated circuit incorporates amplifiers and A/D converters, which lowers the cost for cameras since it contains all the necessary logic to produce an image. CMOS sensors have better chances of integration with other functions. However, adding circuitry within the chip can lead to a risk of disturbances. Moreover, CMOS sensors are much faster with low power consumption and high noise immunity.

6.6.2 Components of a CMOS Sensor

The main parts of a CMOS sensor are: the color filter, the pixels matrix, the digital controller, and the analog to digital converter (Fig. 6.29). Normally, above the pixel matrix is placed a matrix of filters of primary colors to capture the information on the colors due to the incident light. The matrix of pixels consists of millions of sensitive pixels for detecting the light as shown in Fig. 6.30 (checkerboard). The analog signal of the pixels generated by photoelectric effect is sent to the ADC.

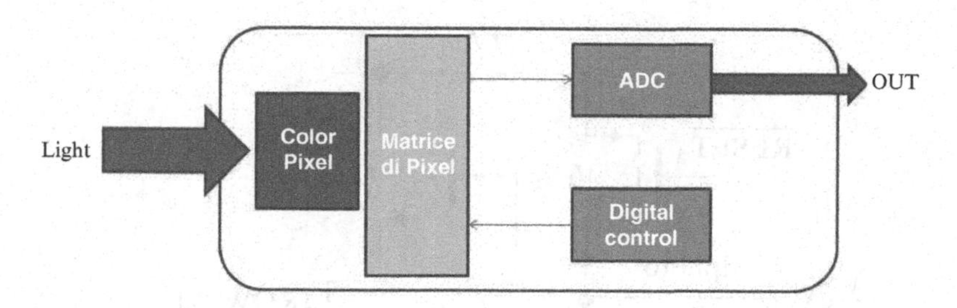

Fig. 6.29 Diagram of a CMOS

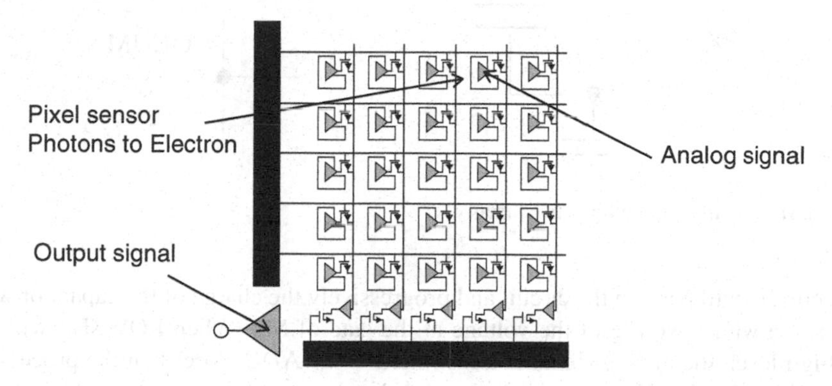

Fig. 6.30 Matrix of pixels

The analog-digital converter (ADC) transforms the output signal of each pixel into a digital signal; then is sent subsequently to an external image processor to obtain the signal viewable. Finally, the digital control manages the pixel matrix. Main components are: clock generator and the oscillator that allow to synchronize events, such as the detection of light [7, 8].

6.6.3 Layout of the Active Pixel

The main parts of a pixel are represented by a photodiode, which detects the photon, and an amplifier. In Fig. 6.31 an example of a circuit constituted by a photodiode, a capacitor C (capacitance resulting between the diode junction and the gate) and a p-MOS transistor M1 and two n-MOS transistor M2 and M3. Each gate of the transistor M1 and M3 is driven by the digital signals RESET and ROWSEL; Vdd is the supply voltage.

The photodiode is reverse-biased to allow passage of electrons during the light detection. The current intensity is almost constant even if the voltage is changing. The photodiode is placed in parallel to a capacitor, that by means of the RESET command is loaded to the supply voltage. During the integration period of the diode,

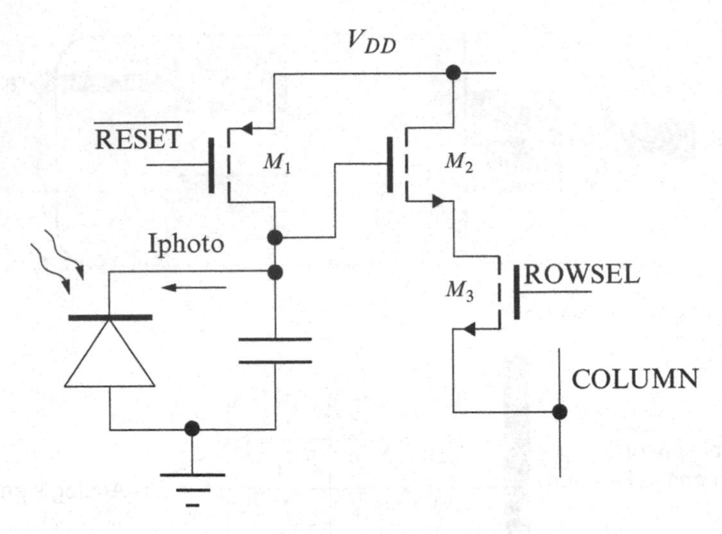

Fig. 6.31 Circuit diagram for active pixels

the current will pass on the circuit and progressively the charge of the capacitor will decrease with lowering of the voltage at the gate of M2. When ROWSEL will be at high level, the analog signal is transmitted to the ADC. After that the process is concluded, ROSWEL returns to the low level and re-start a new cycle of revelation.

6.6.4 Noise Sources

CMOS sensors are affected by dark current of the order of fA at room temperature, which contributes to the discharge of the capacitor and therefore providing an insufficient intensity of the light. Two other forms of noise are the Shot and Flicker (or 1/f noise), related to the fluctuation in the number of carriers. The first can be modeled by a Poisson process and is originated from the discrete nature of electric charge. The Ficker noise occurs in almost all electronic devices, and can show up with a variety of other effects, such as impurities in a conductive channel. It is a consequence of the presence of traps due to crystal defects of the device, which capture or release charge carriers causing the signal fluctuations.

6.6.5 Commercial CMOS

The main companies in the production of CMOS sensors are the Canon, Sony, and Samsung. Samsung offers a wide range of solutions for image sensors for applications such as smartphones, tablets, single-lens reflex (SLR) digital cameras, webcams, automotive cameras, and surveillance cameras. The high performance and energy efficiency are design key for existing mobile applications and digital cameras. The technology Back Side Illumination of Samsung (BSI) has been designed to overcome the degradation of natural light sensitivity and image quality. As a result, Samsung sensors provide high sensitivity for video recording at high speed. Examples of CMOS sensors is the Samsung S5K3H7 for mobile applications with 8MP and 30 fps for HD video recording [8].

6.7 4D Microelectronic

Transistor 4D shaped like Christmas tree could be the next active devices inside a computer. US researchers have created some time ago a new type of transistor and mosfet with a material, gallium–indium arsenide (InGaAs), which could replace the silicon from here to the next 5 years. The 4D electronics is based on previous research where a team of researchers had developed a 3D structure instead of traditional dishes transistor. The technique allows to obtain much faster integrated circuits, compact but especially more efficient in terms of power management, reducing the cost of the system caused by heat sinks and various techniques in order to reduce the temperature of an electronic device. Silicon has an electron mobility limited, indium–gallium arsenide (InGaAs) semiconductor is one of the most promising to replace the silicon due to its excellent features regarding the flow velocity of the electrons. In fact, the silicon is reaching its limits that will prevent to maintain valid Moore's law. The gallium arsenide-indium is already known in optic fiber communications for its excellent physical-electrical characteristics. With the existing 3D transistors, the length of the gates (base, collector, emitter) is about 22 nm made with a process called "molecular beam epitaxy," in which atoms of indium, gallium, and arsenic form a monocrystalline compound. With the new technology it is expected to reduce even more these lengths to around 10 nm by 2018 [2, 8].

6.7.1 Indium Gallium Arsenide

The InGaAs is a chemical compound of indium, gallium, and arsenic. Indium and gallium are both of Group III elements (eg boron), while arsenic is an element of the nitrogen group (group V). As they are in the same group, indium and gallium have similar roles in the chemical bond. InGaAs is a semiconductor with applications in electronics and optoelectronics with properties depending on the percentage of the combination of gallium and indium. The optical and mechanical properties of InGaAs can be various, depending on the ratio of InAs and GaAs. The device is normally InGaAs grown on a substrate of indium phosphide (InP). In order to match the lattice constant of InP and avoid mechanical stresses, is used a particular

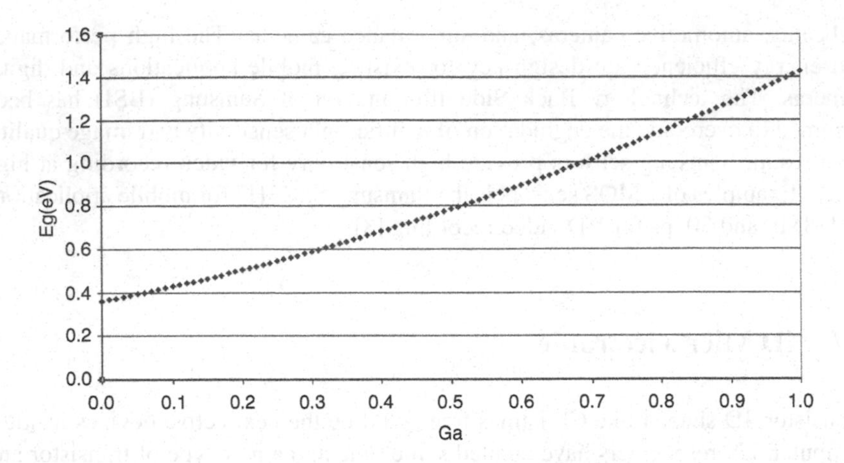

Fig. 6.32 Energy gap as a function of the Gallium composition (Ga) for the semiconductor InGaAs

composition of InGaAs. This composition has a wavelength of cut of 1.68 pM to 295 K. Increasing the molar fraction of InAs compared to GaAs, it is possible to extend the wavelength of cut up to about 2.6 μm. The energy bandgap may be determined from the peak in the spectrum of photoluminescence, provided that the total concentration of impurities is less than 5×10^{16} cm^3. The energy bandgap of the materials depends on the concentration and the temperature. At room temperature is 0.75 eV, between that of germanium and silicon (Fig. 6.32). Coincidentally, the bandgap of InGaAs is in a perfect location for the applications of photodetectors and lasers at a wavelength in the C-band and L and in optical fiber communications. The electron mobility is a fundamental parameter for the design and performance of electronic devices. For the InGaAs is about 10×10^3 cm^2 V^{-1} s^{-1}, which is the largest of any semiconductor, although significantly less than Graphene. The mobility is proportional to the conductivity, a greater mobility shortens the response time of the photodetectors; this improves the efficiency of the device and reduces the power consumption and noise.

6.7.2 Devices

A high electron mobility transistor (HEMT), also known as heterostructure FET (HFET) or MODFET, is a field effect transistor with a junction between two materials with different band gaps (i.e., a heterojunction). InGaAs is one of the new semiconductors for the design of HEMT and for the optical devices (Figs. 6.33 and 6.34). Many features are the high electron mobility, high gain, high output impedance, and transconductance. It's an important device for high speed, high frequency, in digital circuits and microwave circuits, for low-noise applications.

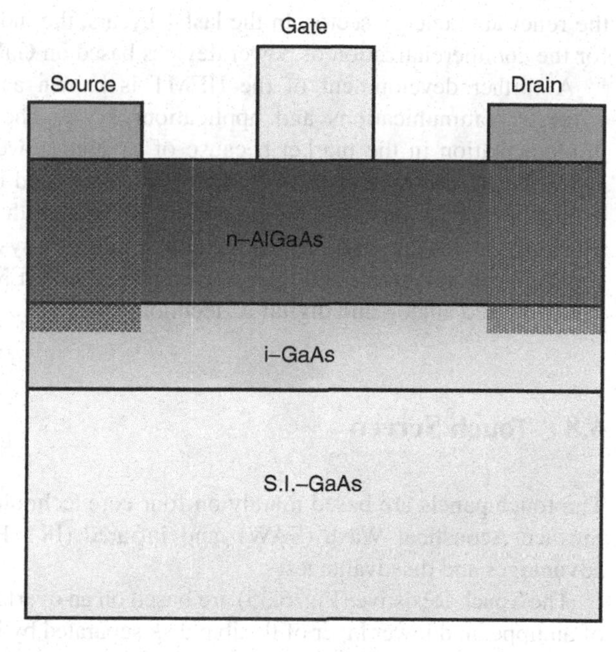

Fig. 6.33 Example of HEMT section

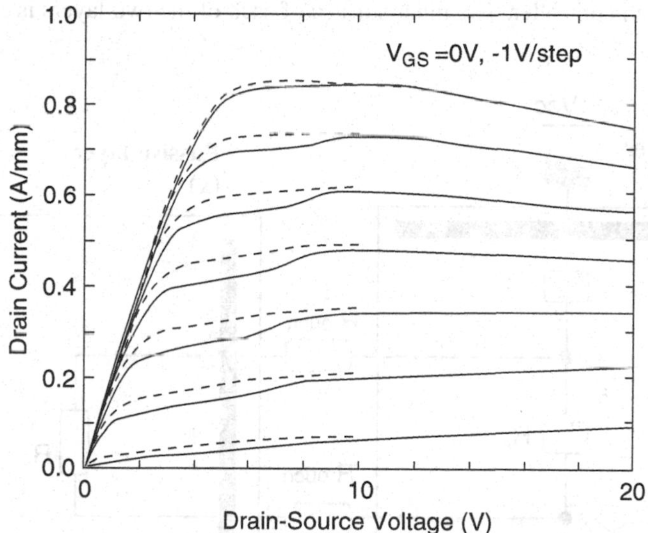

Fig. 6.34 DC characteristic of a HEMT

These applications include the fields of telecommunications, computing, and measurement equipment. HEMT is believed to be a strong candidate for high power switches of the next generation of voltage converters used in many applications of the renewable energy sector. In the last 10 years, the industrial sector has emerged for the commercialization of power devices based on GaN.

A further development of the HEMT is known as PHEMT widely used in wireless communications and applications LNA. The PHEMT transistor finds implementation in the market because of its high power efficiency and excellent low noise figure. As a result, PHEMT are widely used in communication systems by satellite of all forms, including DBS-LNB used with TV and satellite antennas. They are also widely used in satellite communication systems such as general radio communication systems and microwave radar. The PHEMT technology is also used in high-speed analog and digital IC technology [2, 8].

6.8 Touch Screen

The touch panels are based mainly on four core technologies: resistive, capacitive, Surface Acoustical Wave (SAW), and infrared (IR). Each of these designs has advantages and disadvantages.

The Touch Resistive (Fig. 6.35) are based on an overlapping structure, composed of an upper and lower layer of flexible disk separated by insulators and connected to a touch screen controller. The inner surface of each of the two layers is coated with

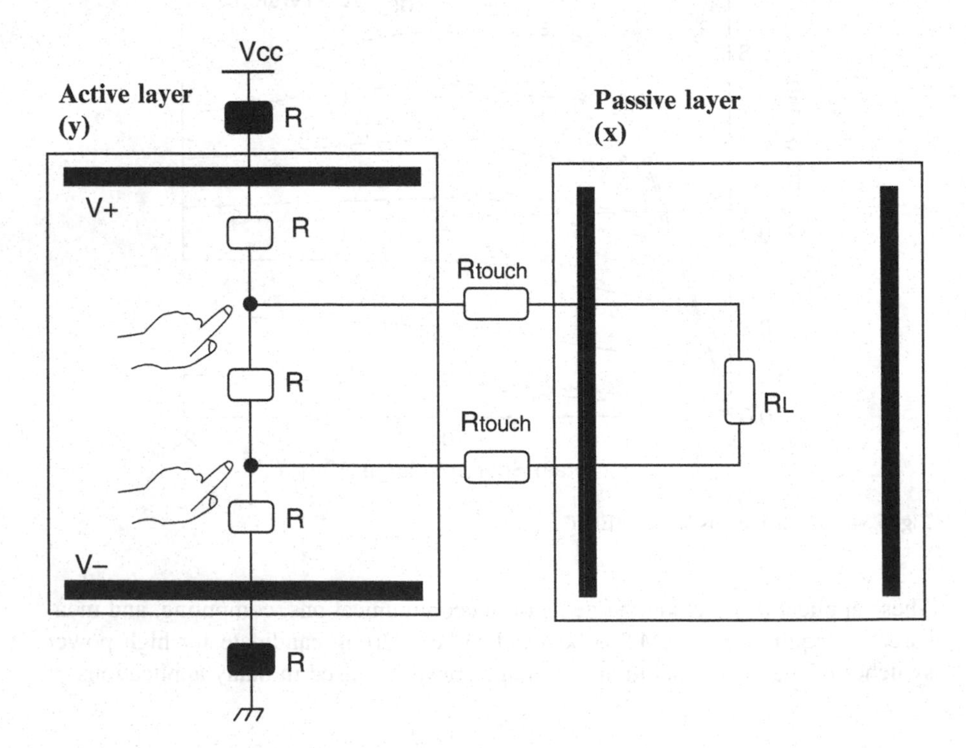

Fig. 6.35 Structure of a resistive touch screen

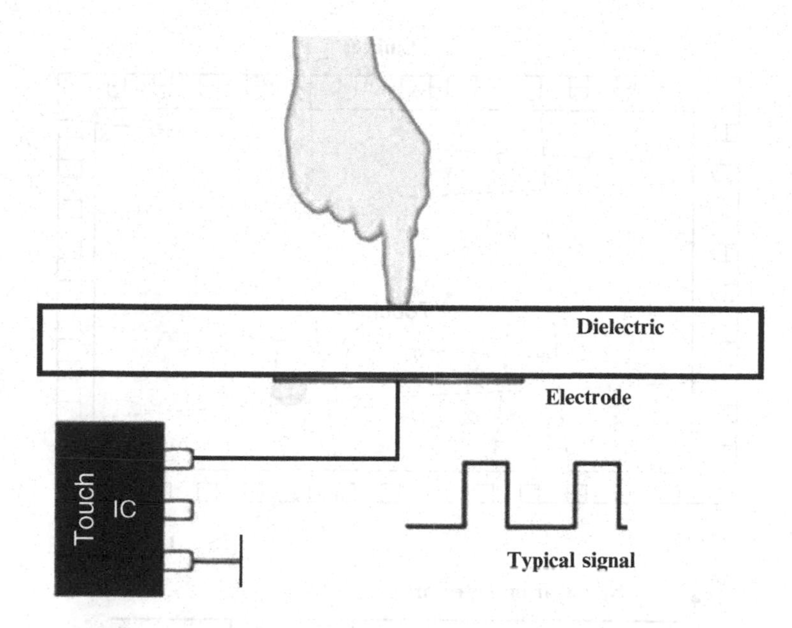

Fig. 6.36 Structure of a capacitive touch screen

a transparent metal oxide, Indium Tin Oxide (ITO), which facilitates the passage of electric charges through each layer when a voltage is applied. Pressing the upper flexible sheet, a current variation due to the electrical contact is produced. The control electronics evaluates the voltage signal and determines the X and Y coordinates resulting from the touch on the screen. The control data are then passed to the operating system of the computer for processing. The resistive touch screen panels are much cheaper and find applications in industrial environments. Main applications of resistive touch screens are in food-service; Point-Of-Sale (POS), industrial process control and instrumentation retail.

A capacitive touch screen panel (Fig. 6.36) is coated with a material similar to the capacitor that has the ability to store the electric charges. When the screen is touched, a small amount of charge is drawn to the contact point. The sensors located on each corner of the panel measure the charge and send the information to the controller for processing. It can be divided into two main categories: surface capacitive and projected. In surface technology, only one side is coated with a conductive layer where a small voltage is applied. When a finger touch is applied, a capacitor is created. The sensor controller can determine the location of touch indirectly by the variation in the capacitance measured by the four corners of the panel. It is used in simple applications such as industrial controls. Technology projected capacitive touch (PCT) consists of a glass layer on top of a touch layer. This layer comprises an XY grid, formed with a set of electrodes separated by two perpendicular layers of conductive material. The touch sensor layer is inserted between the protective cover and the display panel, usually a liquid crystal display

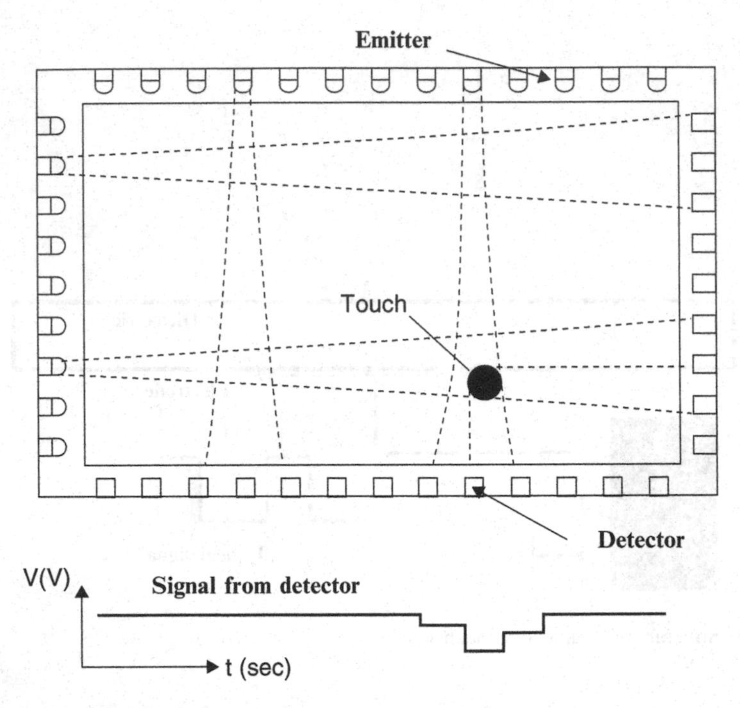

Fig. 6.37 Structure of an IR touch screen

or organic light emitting diodes (OLED). The use of an XY grid allows a higher resolution than resistive technology. The PCT touch screen are more robust and allow the use of gloves. All these features make them ideal for extreme applications, industrial or outdoor [2, 8].

Another technology is the infrared touch screen (Fig. 6.37) that uses an X-Y infrared LED matrix and photodetectors around the edges of the screen to detect the positioning of the touch. A significant advantage is that it can detect essentially any touch input including a finger, finger with the glove, stylus, or pen. Generally it is used in external applications and systems Point-Of-Sale. The SAW touch screen is based on two transducers (transmitter and receiver) both placed on the axes X and Y of the touch panel (Fig. 6.38) and reflectors positioned on the glass. The controller sends the electrical signal to the transducer which converts the signal into ultrasonic waves. Then, the ultrasonic waves are sent to reflectors that are aligned along the edge of the panel. The reflectors diffuse the waves which are collected by another transducer to obtain an electrical signal to be sent to the controller. When a finger touches the screen, the waves are absorbed and appropriately revealed and analyzed by means of control IC. In comparison with resistive and capacitive technologies, SAW technology provides a good image resolution, with good transmission and reception of light. The disadvantages reside in the robustness, the screen SAW is not completely sealed and can be influenced by a large amount of dirt, dust and/or water.

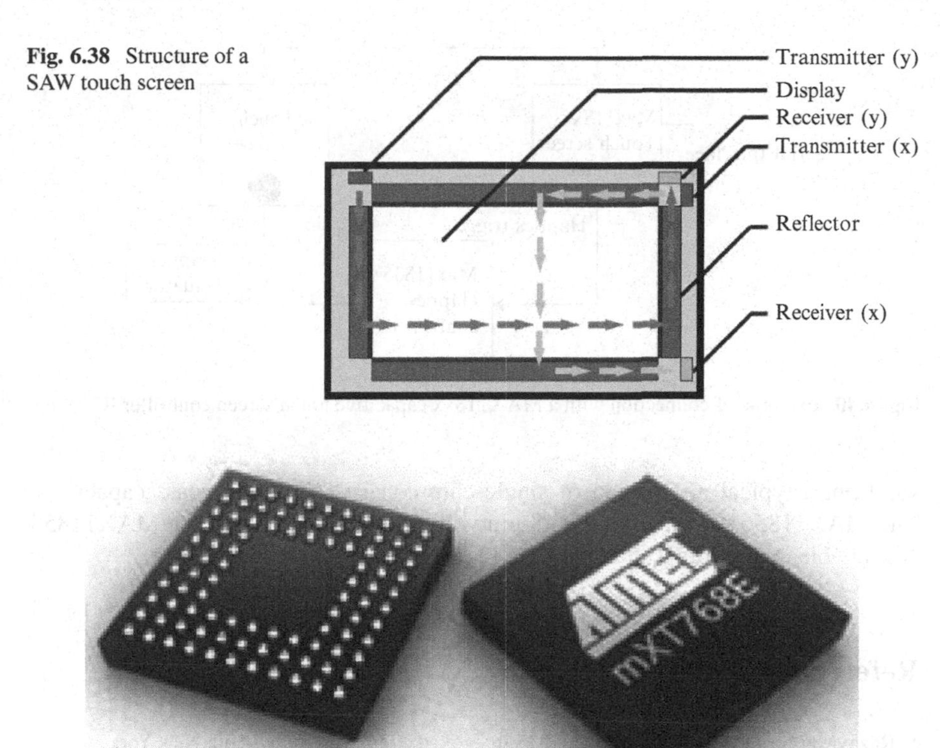

Fig. 6.38 Structure of a SAW touch screen

Fig. 6.39 Atmel maXTouch touch screen controller IC

6.8.1 Controller IC

The CYTT31X controller supports a passive stylus input with a fine point of about 2.5 mm, which is essential for writing languages such as Chinese, Korean, and Japanese. The controller supports CYTT21X a face detection feature that prevents accidental touches.

The ICs of the Atmel (Fig. 6.39), Atmel maXTouch, combine patented technology relating to charge transfer with an Atmel microcontroller AVR 8/16-bit or 32-bit (MCU). These advanced devices provide unlimited touch (up to 16 touches), fast response time, and intelligent processing of an image with multitouch technology for a number of events integrated into the host processor. A higher SNR allows these devices to work well with a finger and stylus touch. The high sensitivity of the MAX11855/MAX11856 capacitive touch screen controllers (Fig. 6.40) of Maxim guarantee low power sensitive applications such as handheld devices. They also have a dedicated input for the suppression of noise events by phone and switching LCD. A minimum number of external components is required to implement a complete

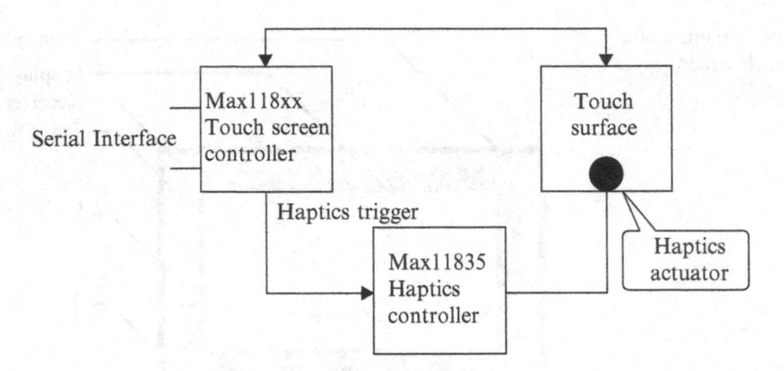

Fig. 6.40 Example of connection with a MAX118xx capacitive touch screen controller IC

solution; a typical application of single-chip consists of three bypass capacitors. The MAX11856 is available in a 48-pin with TQFN package, and the MAX11855 is available in a 40-pin TQFN package [2, 8].

References

1. Razavi B (2002) Design of analog CMOS integrated circuits. McGraw-Hill, New York
2. Di Paolo Emilio M (2013) Data Acquisition system from fundamentals to applied design. Springer, New York
3. Razavi B (2008) Fundamentals of microelectronics. Wiley, Hoboken
4. Neudeck GW (1989) Modular series on solid state devices: volume III: the bipolar junction transistor, 2nd edn. Prentice Hall, Englewood Cliffs
5. Sedra AS, Smith KC (2013) Microelectronic circuits Oxford University Press, New York
6. Razavi B (2002) Design of integrated circuits for optical communications. McGraw-Hill, New York
7. Gray PR, Hurst P, Meyer RG, Lewis S (2001) Analysis and design of analog integrated circuits Wiley, New York
8. Fonstad CG (1994) Microelectronic device and circuits. McGraw-Hill, New York

Index

M. Di Paolo Emilio, *Microelectronics*, DOI 10.1007/978-3-319-22545-6

Printed in the United States
By Bookmasters